CONTENTS

PEARSON EDEXCEL INTERNATIONAL A LEVEL

PHYSICS

Student Book 2

Miles Hudson

Published by Pearson Education Limited, 80 Strand, London, WC2R 0RL.

www.pearsonglobalschools.com

Copies of official specifications for all Pearson Edexcel qualifications may be found on the website: https://qualifications.pearson.com

Text © Pearson Education Limited 2019
Designed by Tech-Set Ltd, Gateshead, UK
Edited by Kate Blackham, Susan Lyons and Jane Read
Typeset by Tech-Set Ltd, Gateshead, UK
Original illustrations © Pearson Education Limited 2019
Cover design by Pearson Education Limited 2019
Picture research by Integra
Cover photo © Visuals Unlimited, Inc./Dr. Robert Gendler
Inside front cover photo: Dmitry Lobanov

The right of Miles Hudson to be identified as author of this work has been asserted by him in accordance with the Copyright, Designs and Patents Act 1988.

First published 2019

25 24
10

British Library Cataloguing in Publication Data
A catalogue record for this book is available from the British Library
ISBN 978 1 2922 4447 8

Printed in Slovakia by Neografia

Endorsement statement
In order to ensure that this resource offers high-quality support for the associated Pearson qualification, it has been through a review process by the awarding body. This process confirmed that this resource fully covers the teaching and learning content of the specification at which it is aimed. It also confirms that it demonstrates an appropriate balance between the development of subject skills, knowledge and understanding, in addition to preparation for assessment.

Endorsement does not cover any guidance on assessment activities or processes (e.g. practice questions or advice on how to answer assessment questions) included in the resource, nor does it prescribe any particular approach to the teaching or delivery of a related course.

While the publishers have made every attempt to ensure that advice on the qualification and its assessment is accurate, the official specification and associated assessment guidance materials are the only authoritative source of information and should always be referred to for definitive guidance.

Pearson examiners have not contributed to any sections in this resource relevant to examination papers for which they have responsibility.

Examiners will not use endorsed resources as a source of material for any assessment set by Pearson. Endorsement of a resource does not mean that the resource is required to achieve this Pearson qualification, nor does it mean that it is the only suitable material available to support the qualification, and any resource lists produced by the awarding body shall include this and other appropriate resources.

Text Credits:
10 The Planetary Society: Kim Orr/ The Planetary Society **22 Theodore W. Hall:** From a webpage to calculate artificial gravity, maintained by Ted Hall: Hall, Theodore W. (2012). 'SpinCalc: An Artificial-Gravity Calculator in JavaScript'; www.artificial-gravity.com/sw/SpinCalc/SpinCalc.htm (as at 24 October 2014). **22 National Space Society:** Space Settlement Population Rotation Tolerance, Al Globus, Theodore Hall, June 2015, © National Space Society 2015 **22 NASA:** 6th NASA Symposium on The Role of the Vestibular Organs in the Exploration of Space, Portland, OR, USA, September 30–October 3, 2002, Journal of Vestibular Research, vol. 13, no. 4-6, pp. 321-330, 2003 © IOS Press, Inc. 2003. **38 Bonneville Power Administration:** Section 3.1 from Appendix E, Electrical Effects, of 'BIG EDDY – KNIGHT 500-kV TRANSMISSION PROJECT', by T. Dan Bracken, for Bonneville Power Administration, dated March 2010, https://www.bpa.gov/efw/Analysis/NEPADocuments/nepa/Big_Eddy-Knight/AppendixE-ElectricalEffectsCombined.pdf **52 AIP Publishing:** Reproduced from 'Paper-based ultracapacitors with carbon nanotubes-graphene composites', Journal of Applied Physics, Vol. 115, Issue 16 (Li, J., Cheng, X., Sun, J., Brand, C., Shashurin, A., Reeves, M. and Keidar, M. 2014), with the permission of AIP Publishing **68 Minelab Electronics:** From a paper written by Bruce Candy, Chief Scientist, Minelab Electronics, a manufacturer of metal detectors. https://www.minelab.com/__files/f/11043/METAL DETECTOR BASICS AND THEORY.pdf. **80 Guardian News and Media Limited:** © Guardian News and Media Limited, 2018 **96 IOP Publishing:** Extract from an online article by Tami Freeman, editor of medicalphysicsweb posted on 24 February 2014 at http://medicalphysicsweb.org/cws/article/opinion/56295. © IOP Publishing. Reproduced with permission. All rights reserved. **112 NASA:** Extract from the NASA website, at http://www.nasa.gov/centers/glenn/technology/warp/antistat.html posted on 2 May 2008. The article has since been updated with a focus on warp-drive technology. **128 National Geographic Society:** Bisharat, Andrew, After Crossing Pacific, Record-Setting Balloonists Land Off Baja Coast, for National Geographic, Jan 31, 2015, © National Geographic Partners, LLC, 2015 **146 American Institute of Physics:** Gwynne, Peter, Medical Imaging Faces Shortage of Key Radioactive Material, Inside Science, Nov 8, 2016. © American Institute of Physics 2016 **164 Taipei Financial Center Corp:** Best in the World : Wind Damper, TAIPEI 101 © Taipei Financial Center Corp. 2014 **164 Amusing Planet:** Patowary, Kaushik, The 728-Ton Tuned Mass Damper of Taipei 101 © Amusing Planet 2014 **174 Dunedin Academic Press:** From pages 86–87 and page 89 (in Chapter 9) of Introducing Volcanology by Dr Dougal Jerram, ISBN 978-1-90671-622-6, published by Dunedin Academic Press (2011). **200 Michael O Neill:** Michael O Neill, © 2018 The Universe for Kids

ABOUT THIS BOOK

This book is written for students following the Pearson Edexcel International Advanced Level (IAL) Physics specification. This book covers the second year of the International A Level course.

The book contains full coverage of IAL units (or exam papers) 4 and 5. Unit 4 in the specification has three topic areas; Unit 5 has four topic areas. The topics in this book, and their contents, fully match the specification. You can refer to the Assessment Overview on pages x–xi for further information. Students can prepare for the written Practical Skills Paper (Unit 6) with the support of the IAL Physics Lab Book (see pages viii and ix of this book).

Each Topic is divided into chapters and sections so that the content is presented in manageable chunks. Each section features a mix of learning and activities supported by features explained below.

Learning objectives
Each chapter starts with a list of key learning objectives.

Specification reference
The exact specification references covered in the section are listed.

Exam hints
Tips on how to answer exam-style questions and guidance for exam preparation, including how to respond to **command words**. Content which you do not need to revise for your exams is indicated by red **Exam Hint: Extra content** boxes.

Worked examples
These show you how to work through questions and set out calculations.

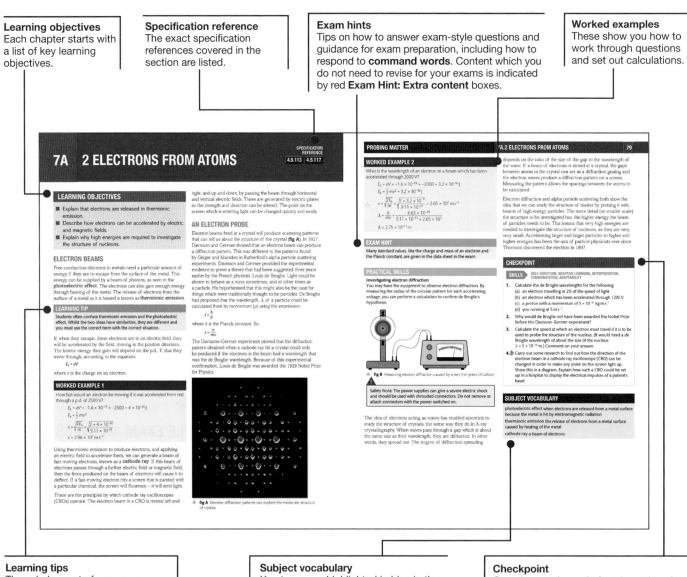

Learning tips
These help you to focus your learning and avoid common errors.

Subject vocabulary
Key terms are highlighted in blue in the text. Clear definitions are provided at the end of each section for easy reference, and are also collated in the **glossary** at the back of the book.

Checkpoint
Questions at the end of each section check understanding of the key learning points. Certain questions allow you to develop **skills** which will be valuable for further study and in the workplace.

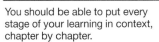

You should be able to put every stage of your learning in context, chapter by chapter.

- Links to other areas of Physics include previous knowledge that is built on in the chapter, and areas of knowledge and application that you will cover later in your course.
- Maths knowledge required is detailed in a handy checklist. If you need to practise the maths you need, you can use the **Maths Skills** reference at the back of the book as a starting point.

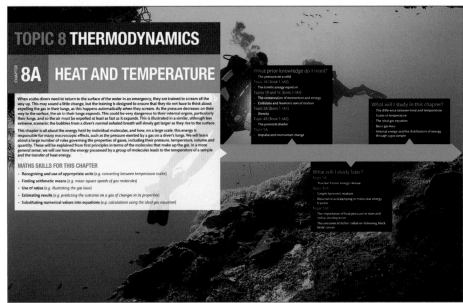

Thinking Bigger

At the end of each topic, there is an opportunity to read and work with real-life research and writing about science.

The activities help you to read authentic material that's relevant to your course, analyse how scientists write, think critically and consider how different aspects of your learning piece together.

These Thinking Bigger activities focus on **key transferable skills**, which are an important basis for key academic qualities.

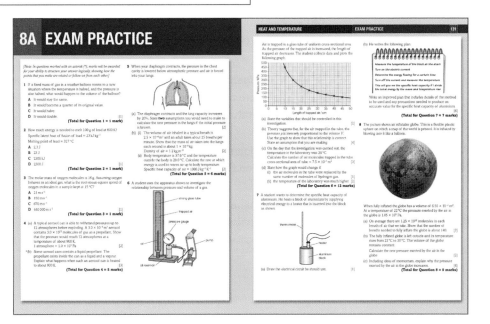

Exam Practice

Exam-style questions at the end of each chapter are tailored to the Pearson Edexcel specification to allow for practice and development of exam writing technique. They also allow for practice responding to the "command words" used in the exams (see the **command words glossary** at the back of this book).

The **Preparing for your exams** section at the end of the book includes **sample answers** for different question types, with comments about the strengths and weaknesses of the answers.

PRACTICAL SKILLS

Practical work is central to the study of physics. The second year of the Pearson Edexcel International Advanced Level (IAL) Physics course includes eight Core Practicals that link theoretical knowledge and understanding to practical scenarios.

Your knowledge and understanding of practical skills and activities will be assessed in all exam papers for the IAL Physics qualification.

- Papers 4 and 5 will include questions based on practical activities, including novel scenarios.
- Paper 6 will test your ability to plan practical work, including risk management and selection of apparatus.

In order to develop practical skills, you should carry out a range of practical experiments related to the topics covered in your course. Further suggestions in addition to the Core Practicals are included below.

STUDENT BOOK TOPIC	IAL CORE PRACTICALS	
TOPIC 5 **FURTHER MECHANICS**	**CP9**	Investigate the relationship between the force exerted on an object and its change of momentum
	CP10	Use ICT to analyse collisions between small spheres
TOPIC 6 **ELECTRIC AND MAGNETIC FIELDS**	**CP11**	Use an oscilloscope or data logger to display and analyse the potential difference (p.d.) across a capacitor as it charges and discharges through a resistor
TOPIC 8 **THERMODYNAMICS**	**CP12**	Calibrate a thermistor in a potential divider circuit as a thermostat
	CP13	Determine the specific latent heat of a phase change
	CP14	Investigate the relationship between pressure and volume of a gas at fixed temperature
TOPIC 9 **NUCLEAR DECAY**	**CP15**	Investigate the absorption of gamma radiation by lead
TOPIC 10 **OSCILLATIONS**	**CP16**	Determine the value of an unknown mass using the resonant frequencies of the oscillation of known masses

UNIT 4 (TOPICS 5 TO 7)
FURTHER MECHANICS, FIELDS AND PARTICLES

Possible further practicals include:

- investigating the effect of mass, velocity and radius of orbit on centripetal force
- using a coulomb meter to measure charge stored
- using an electronic balance to measure the force between two charges

UNIT 5 (TOPICS 8 TO 11)
THERMODYNAMICS, RADIATION, OSCILLATIONS AND COSMOLOGY

Possible further practicals include:

- investigating the relationship between the volume and temperature of a fixed mass of gas
- measuring the half-life of a radioactive material, measuring gravitational field strength using a simple pendulum and measuring a spring constant from simple harmonic motion

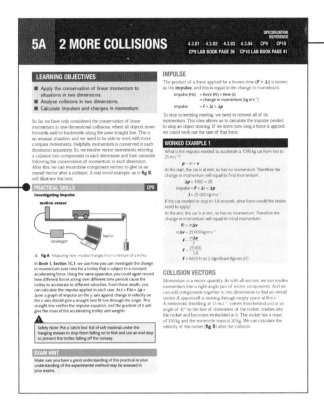

In the **Student Book**, the Core Practical specification and Lab Book references are supplied in the relevant sections.

5A 2 MORE COLLISIONS

SPECIFICATION REFERENCE
4.3.81 | 4.3.82 | 4.3.83 | 4.3.84 | CP9 | CP10
CP9 LAB BOOK PAGE 36 | CP10 LAB BOOK PAGE 41

LEARNING OBJECTIVES

■ Apply the conservation of linear momentum to situations in two dimensions.
■ Analyse collisions in two dimensions.
■ Calculate impulses and changes in momentum.

So far, we have only considered the conservation of linear momentum in one-dimensional collisions, where all objects move forwards and/or backwards along the *same* straight line. This is an unusual situation, and we need to be able to work with more complex movements. Helpfully, momentum is conserved in each dimension separately. So, we resolve vector movements entering a collision into components in each dimension and then calculate following the conservation of momentum in each dimension. After this, we can recombine component vectors to give us an overall vector after a collision, as in **fig B**, will illustrate this best.

PRACTICAL SKILLS   CP9

Investigating impulse

motion sensor

datalogger   *laptop*

▲ **fig A** Measuring how impulse changes the momentum of a trolley

In Book 1, Section 1C.1, we saw how you can investigate the change in momentum over time for a trolley that is subject to a constant accelerating force. Using the same apparatus, you could again record how different forces acting over different time periods cause the trolley to accelerate to different velocities. From these results, you can calculate the impulse applied in each case. As $I = F\Delta t = \Delta p = \Delta mv$, a graph of impulse on the y-axis against change in velocity on the x-axis should give a straight best fit line through the origin. This straight line verifies the impulse equation, and the gradient of it will give the mass of the accelerating trolley and weights.

⚠ Safety Note: Put a 'catch box' full of soft material under the hanging masses to stop them falling on to feet and use an end stop to prevent the trolley falling off the runway.

EXAM HINT

Make sure you have a good understanding of this practical as your understanding of the experimental method may be assessed in your exams.

IMPULSE

The product of a force applied for a known time ($F \times \Delta t$) is known as the **impulse**, and this is equal to the change in momentum.

impulse (Ns) = force (N) × time (s)
              = change in momentum (kg m s^{-1})
impulse   = $F \times \Delta t = \Delta p$

To stop something moving, we need to remove all of its momentum. This idea allows us to calculate the impulse needed to stop an object moving. If we know how long a force is applied, we could work out the size of that force.

WORKED EXAMPLE 1

What is the impulse needed to accelerate a 1000 kg car from rest to 25 m s^{-1}?

$$p = m \times v$$

At the start, the car is at rest, so has no momentum. Therefore the change in momentum will equal its final momentum.

$$\Delta p = 1000 \times 25$$
$$impulse = F \times \Delta t = \Delta p$$
$$I = 25\,000 \text{ kg m s}^{-1}$$

If the car needed to stop in 3.8 seconds, what force would the brakes need to apply?

At the end, the car is at rest, so has no momentum. Therefore the change in momentum will equal its initial momentum.

$$Ft = m\Delta v$$
$$m\Delta v = 25\,000 \text{ kg m s}^{-1}$$
$$F = \frac{m\Delta v}{t}$$
$$F = \frac{25\,000}{3.8}$$
$$F = 6600 \text{ N to 2 significant figures (sf)}$$

COLLISION VECTORS

Momentum is a vector quantity. As with all vectors, we can resolve momentum into a right-angle pair of vector components. And we can add components together in two dimensions to find an overall vector. A spacecraft is moving through empty space at 8 m s^{-1}. A meteoroid, travelling at 15 m s^{-1}, comes from behind and at an angle of 45° to the line of movement of the rocket, crashes into the rocket and becomes embedded in it. The rocket has a mass of 350 kg and the meteorite mass is 20 kg. We can calculate the velocity of the rocket (**fig B**) after the collision.

Practical Skills

Practical skills boxes explain techniques or apparatus used in the Core Practicals, and also detail useful skills and knowledge gained in other related investigations.

CORE PRACTICAL 9:
INVESTIGATE THE RELATIONSHIP BETWEEN THE FORCE EXERTED ON AN OBJECT AND ITS CHANGE OF MOMENTUM

SPECIFICATION REFERENCE
4.3.82

Procedure

1 Secure the bench pulley to one end of the runway. This end of the runway should project over the end of a bench, so that the string connecting the mass hanger and the trolley passes over the pulley. The mass hanger will fall to the floor as the trolley moves along the runway. The runway should be tilted to compensate for friction.

2 Place the slotted mass hanger on the floor and move the trolley backwards along the runway until the string becomes tight, with the mass still on the floor. Place the light gate so it is positioned in the middle of the interrupt card on the trolley. There should be enough space on the ramp to allow the trolley to continue until it clears the light gate before hitting the pulley.

3 Move the trolley further backwards until the mass hanger is touching the pulley. Put the five 10g masses on the trolley so that they will not slide off. This is the start position for the experiment.

4 Record the total hanging mass, m. Release the trolley and use the stop clock to measure the time, T, it takes for the trolley to move from the start position to the light gate – this should be when the mass hanger hits the floor. Record the time reading, t, on the light gate. Repeat your measurements twice more and calculate mean values for T and t. Then estimate δT and δt, the uncertainties in these values.

5 Move one 10g mass from the trolley to the hanger and repeat step 4. Repeat this process, moving one 10g mass at a time and recording m, T and t; until all of the masses are on the hanger.

6 Measure the combined mass, M, of the trolley, string, slotted masses and hanger.

7 Measure the distance, d, travelled by the trolley. This should be the same as the distance fallen by the mass hanger.

8 Record the length, L, of the card.

9 You can develop the investigation further by taking more readings after adding an additional mass, for example, 200 g, to the mass of the trolley.

Learning tip

● Choose a suitable scale for your graph so that your plot fills the whole page – you do not need to include the origin. This will make it easier to draw the last two gradient lines.

Objectives

● To determine the momentum change of a trolley when a force acts on it, as a function of time

Equipment

● five slotted masses (10g) and hanger
● light gate and recorder
● stop clock
● metre rule
● dynamics trolley or air track vehicle
● runway or air track
● bench pulley
● string

⚠ **Safety**

● Runways and trolleys are very heavy and need to be placed so they will not slide or fall off benches.
● Air track blowers should be on the floor with the hose secured so that it cannot come loose and blow dust and dirt into people's faces.
● If large masses are used a catch box is needed in the drop zone to keep feet clear.

CORE PRACTICAL 9:
INVESTIGATE THE RELATIONSHIP BETWEEN THE FORCE EXERTED ON AN OBJECT AND ITS CHANGE OF MOMENTUM

SPECIFICATION REFERENCE
4.3.82

2 Find the gradient of your line of best fit and compare it with your value for $\frac{M}{g}$

3 You can take the uncertainty in T and t as half the range of repeated readings. You need not work out the uncertainty for every value of T and t, but take typical values, neither the largest nor the smallest.

a Calculate δv, the actual uncertainty in v, from the equation $\delta v = v\left(\frac{\delta t}{t}\right)$. Use a mid-range value for v.

b Calculate $\delta(mT)$ by multiplying a mid-range value for m (for example, 30g) by δT.

c Use these actual uncertainties to draw error bars in both directions to form error boxes on your graph. Draw one line that is steeper than the line of best fit (LoBF) and one line that is less steep than the LoBF. Both of these lines should pass through the error boxes. Find the gradient of each new line.

The difference between the two gradients of the lines gives you the uncertainty in your gradient and this uncertainty is based on your readings.

Your value for $\frac{M}{g}$ should lie between these two values if Newton's second law is operating.

This Student Book is accompanied by a **Lab Book**, which includes instructions and writing frames for the Core Practicals for students to record their results and reflect on their work.

Practical skills checklists, practice questions and answers are also provided.

The Lab Book records can be used as preparation and revision for the Practical Skills Papers.

ASSESSMENT OVERVIEW

The following tables give an overview of the assessment for the second year of the Pearson Edexcel International Advanced Level course in Physics. You should study this information closely to help ensure that you are fully prepared for this course and know exactly what to expect in each part of the exam. More information about this qualification, and about the question types in the different papers, can be found on page 210 of this book.

PAPER / UNIT 4	PERCENTAGE OF IA2	PERCENTAGE OF IAL	MARK	TIME	AVAILABILITY
FURTHER MECHANICS, FIELDS AND PARTICLES Written exam paper Paper code WPH14/01 Externally set and marked by Pearson Edexcel Single tier of entry	40%	20%	90	1 hour 45 minutes	January, June and October First assessment: January 2020

PAPER / UNIT 5	PERCENTAGE OF IA2	PERCENTAGE OF IAL	MARK	TIME	AVAILABILITY
THERMODYNAMICS, RADIATION, OSCILLATIONS AND COSMOLOGY Written exam paper Paper code WPH15/01 Externally set and marked by Pearson Edexcel Single tier of entry	40%	20%	90	1 hour 45 minutes	January, June and October First assessment: June 2020

PAPER / UNIT 6	PERCENTAGE OF IA2	PERCENTAGE OF IAL	MARK	TIME	AVAILABILITY
PRACTICAL SKILLS IN PHYSICS II Written exam paper Paper code WPH16/01 Externally set and marked by Pearson Edexcel Single tier of entry	20%	10%	50	1 hour 20 minutes	January, June and October First assessment: June 2020

ASSESSMENT OBJECTIVES AND WEIGHTINGS

ASSESSMENT OBJECTIVE	DESCRIPTION	% IN IAS	% IN IA2	% IN IAL
A01	Demonstrate knowledge and understanding of science	34–36	29–31	32–34
A02	(a) Application of knowledge and understanding of science in familiar and unfamiliar contexts	34–36	33–36	34–36
	(b) Analysis and evaluation of scientific information to make judgments and reach conclusions	9–11	14–16	11–14
A03	Experimental skills in science, including analysis and evaluation of data and methods	20	20	20

RELATIONSHIP OF ASSESSMENT OBJECTIVES TO UNITS

UNIT NUMBER	ASSESSMENT OBJECTIVE (%)			
	A01	A02 (A)	A02 (B)	A03
UNIT 1	17–18	17–18	4.5–5.5	0
UNIT 2	17–18	17–18	4.5–5.5	0
UNIT 3	0	0	0	20
TOTAL FOR INTERNATIONAL ADVANCED SUBSIDIARY	33–36	34–36	9–11	20

UNIT NUMBER	ASSESSMENT OBJECTIVE (%)			
	A01	A02 (A)	A02 (B)	A03
UNIT 1	8.5–9.0	8.5–9.0	2.25–2.75	0
UNIT 2	8.5–9.0	8.5–9.0	2.25–2.75	0
UNIT 3	0	0	0	10
UNIT 4	7.3–7.8	8.4–8.9	3.6–4.0	0
UNIT 5	7.3–7.8	8.4–8.9	3.6–4.0	0
UNIT 6	0	0	0	10
TOTAL FOR INTERNATIONAL ADVANCED LEVEL	32–34	34–36	11–14	20

TOPIC 5 FURTHER MECHANICS

5A FURTHER MOMENTUM

Acceleration can be considered as a change in momentum per unit mass. It can often be more exciting than basic calculations such as calculating the changing speed of a car.

Curling is a game which originated in Scotland. It is now an established sport which is popular in Canada and Japan. The sport uses the ideas of conservation of momentum and elastic collisions. Players deliberately collide the stones to deflect their opponents' stones, and to ensure their own stone finishes in a winning position. Also, the friction with the ice causes a change in momentum to slow the stone to a stop.

In this chapter, you will learn about the way forces can change the momentum of an object over time. The chapter will also cover how kinetic energy changes in different types of collisions, whilst momentum is conserved. All of this will be extended to events happening in two dimensions, so vector addition and the resolving of vectors will be revisited in order to make the necessary calculations.

MATHS SKILLS FOR THIS CHAPTER

- **Use of trigonometric functions** (*e.g. finding components of momentum vectors*)
- **Use of Pythagoras' theorem** (*e.g. finding velocity as a vector sum*)
- **Changing the subject of an equation** (*e.g. rearranging the impulse equation*)
- **Substituting numerical values into algebraic equations** (*e.g. finding the velocity after a collision*)
- **Visualising and representing 2D forms** (*e.g. drawing a 2D momentum diagram for a collision between meteors*)

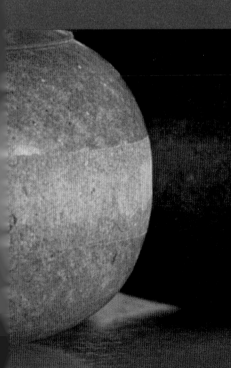

What prior knowledge do I need?

Topic 1A (Book 1: IAS)

- How to add forces as vectors
- How to resolve vectors
- Newton's laws of motion

Topic 1B (Book 1: IAS)

- How to calculate kinetic energy
- Conservation of energy

Topic 1C (Book 1: IAS)

- How to calculate the momentum of an object
- Conservation of linear momentum in collisions

What will I study in this chapter?

- The impulse equation and its connection with Newton's second law of motion
- The relationship between the force on an object and its change in momentum
- Conservation of linear momentum in two dimensions
- How to analyse collisions in 2D
- The difference between an elastic and an inelastic collision
- The equation for the kinetic energy of a non-relativistic particle, in terms of its momentum

What will I study later?

Topics 6A and 6B

- How electrical and magnetic fields affect the momentum of charged particles

Topic 7A

- The de Broglie wavelength for a particle and its connection with the momentum of the particle
- Large-angle alpha particle scattering indicating the structure of the atom, with the scattering dependent on momentum conservation

Topic 7B

- How the momentum affects the size of a circle in which a charged particle is trapped by a magnetic field
- How conservation of momentum affects the creation and detection of new particles

■ Explain the difference between elastic and inelastic collisions.
■ Make calculations based on the conservation of linear momentum to determine energy changes in collisions.
■ Derive and use the equation for the kinetic energy of a non-relativistic particle.

We have seen in **Book 1, Chapter 1C** that linear momentum is always conserved in any collision between objects, and this is responsible for Newton's third law of motion. We also learned that Newton's second law of motion expresses the concept that a force is equivalent to the rate of change of momentum. **Book 1, Chapter 1A** explained how forces can do work, which results in energy transfer. So, does the kinetic energy change in a collision?

▲ **fig A** Damaging a car uses energy. What can we say about the conservation of kinetic energy in a car crash?

ELASTIC COLLISIONS

In a collision between one pool ball and another, the first one often stops completely and the second then moves away from the collision. As both pool balls have the same mass, the principle of conservation of momentum tells us that the velocity of the second ball must be identical to the initial velocity of the first. This means that the kinetic energy of this system of two balls before and after the collision must be the same. A collision in which kinetic energy is conserved is called an **elastic collision**. In general, these are rare. A Newton's cradle is an example that is nearly perfectly elastic (a tiny amount of energy is lost as heat and sound). A collision caused by non-contact forces, such as alpha particles being scattered by a nucleus (see **Section 7A.1**), is perfectly elastic.

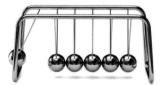

▲ **fig B** Newton's cradle maintains kinetic energy, as well as conserving momentum in its collisions.

INELASTIC COLLISIONS

In a crash between two bumper cars, the total momentum after the collision must be identical to the total momentum before the collision. However, if we calculate the total kinetic energy before and after, we find that the total is reduced by the collision. Some of the kinetic energy is transferred into other forms such as heat and sound. A collision in which total kinetic energy is not conserved is called an **inelastic collision**.

INELASTIC COLLISION EXAMPLE

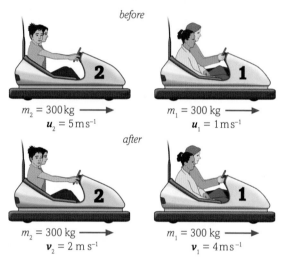

▲ **fig C** The fun of inelastic collisions.

If you calculate the total momentum before and after the collision in **fig C**, you will see that it is conserved. However, what happens to the kinetic energy?

Before collision:
$$E_{k1} = \tfrac{1}{2} m_1 u_1^2 = \tfrac{1}{2} \times (300) \times 1^2 = 150\,J$$
$$E_{k2} = \tfrac{1}{2} m_2 u_2^2 = \tfrac{1}{2} \times (300) \times 5^2 = 3750\,J$$
Total kinetic energy = 3900 J

After collision:
$$E_{k1} = \tfrac{1}{2} m_1 v_1^2 = \tfrac{1}{2} \times (300) \times 4^2 = 2400\,J$$
$$E_{k2} = \tfrac{1}{2} m_2 v_2^2 = \tfrac{1}{2} \times (300) \times 2^2 = 600\,J$$
Total kinetic energy = 3000 J

Loss in kinetic energy = 900 J. This is an inelastic collision.

This 'lost' energy has been transferred to heat and sound energy.

LEARNING TIP

When you are deciding whether a collision is elastic or inelastic, you must only consider the conservation of *kinetic* energy. Total energy in all forms must always be conserved.

PRACTICAL SKILLS

Investigating elastic and inelastic collisions

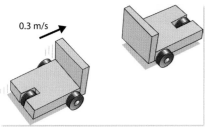
0.3 m/s

▲ **fig D** Crash testing the elasticity of collisions.

You can investigate elastic and inelastic collisions in the school laboratory. If you cause head-on collisions, and record the mass and velocity of each trolley before and after the collisions, you then calculate the momentum at each stage. This should be conserved. You can also then calculate kinetic energy before and after the collisions. Real cars are designed with crumple zones to absorb as much kinetic energy as possible when they crash. This reduces the energy available to cause injury to the passengers. What is the best design for a crumple zone on your experimental trolleys which will absorb kinetic energy?

Safety Note: Carry and place heavy runways so they cannot fall. Use end-stops to prevent the trolleys falling off the ends of the runway.

PARTICLE MOMENTUM

We know that the formula for calculating kinetic energy is $E_k = \frac{1}{2}mv^2$ and that the formula for momentum is $p = mv$. We can combine these to get an equation that gives kinetic energy in terms of the momentum and mass.

$$E_k = \frac{1}{2}mv^2 \quad \text{and} \quad v = \frac{p}{m}$$

$$E_k = \frac{1}{2}mv^2$$

$$\therefore \quad E_k = \frac{1}{2}m\left(\frac{p}{m}\right)^2$$

$$\therefore \quad E_k = \frac{1}{2}\frac{p^2}{m}$$

$$\therefore \quad E_k = \frac{p^2}{2m}$$

This formula is particularly useful for calculations involving the kinetic energy of subatomic particles travelling at non-relativistic speeds – that is, much slower than the speed of light.

PARTICLE COLLISIONS

In experiments to determine the nature of fundamental particles, physicists detect the movements of many unknown particles. The Large Hadron Collider experiment at CERN, underground near Geneva in Switzerland, produces 600 million particle interactions in its detector every second. The conservation of momentum allows the mass of these particles to be calculated, which helps to identify them. This can be done by colliding the particles produced in the experiment with known particles in the detector.

For example, the detector registers an elastic collision with one of its neutrons which changes the neutron's velocity from stationary to $3.4 \times 10^6 \, \text{m s}^{-1}$. The collision was 'head-on' with an unknown particle, which was initially moving at 10% of the speed of light, and leaves the collision in the opposite direction at $1.09 \times 10^3 \, \text{m s}^{-1}$. What is the mass of the mystery particle? The mass of a neutron is $1.67 \times 10^{-27} \, \text{kg}$.

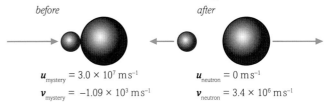
before   *after*

$u_{\text{mystery}} = 3.0 \times 10^7 \, \text{m s}^{-1}$ $u_{\text{neutron}} = 0 \, \text{m s}^{-1}$
$v_{\text{mystery}} = -1.09 \times 10^3 \, \text{m s}^{-1}$ $v_{\text{neutron}} = 3.4 \times 10^6 \, \text{m s}^{-1}$

▲ **fig E** Discovering mystery particles from their momentum and collisions.

Before collision:

$$p_{\text{mystery}} = m_{\text{mystery}} \times 3 \times 10^7 = p_{\text{total before}} \qquad (p_n = \text{zero})$$

After collision:

$$p_{\text{total after}} = (m_{\text{mystery}} \times v_{\text{mystery}}) + (m_n \times v_n)$$
$$= (m_{\text{mystery}} \times -1.09 \times 10^3) + (1.67 \times 10^{-27} \times 3.4 \times 10^6)$$
$$= p_{\text{total before}} = m_{\text{mystery}} \times 3 \times 10^7$$

So:

$$(m_{\text{mystery}} \times -1.09 \times 10^3) + (1.67 \times 10^{-27} \times 3.4 \times 10^6) = m_{\text{mystery}} \times 3 \times 10^7$$
$$(1.67 \times 10^{-27} \times 3.4 \times 10^6) = (m_{\text{mystery}} \times 3 \times 10^7) - (m_{\text{mystery}} \times -1.09 \times 10^3)$$
$$5.678 \times 10^{-21} = (m_{\text{mystery}} \times 30\,001\,090)$$

So:

$$m_{\text{mystery}} = \frac{5.678 \times 10^{-21}}{30\,001\,090} = 1.89 \times 10^{-28} \, \text{kg}$$

This is approximately 207 times the mass of an electron, and so this can be identified as a particle called a muon, which is known to have this mass.

CHECKPOINT

SKILLS ANALYSIS

1. An alpha particle consists of two protons and two neutrons. Calculate the kinetic energy of an alpha particle which has a momentum of $1.08 \times 10^{-19} \, \text{kg m s}^{-1}$:
   (a) in joules   (b) in electron volts   (c) in MeV.
   (mass of neutron = mass of proton = $1.67 \times 10^{-27} \, \text{kg}$)

2. A bowling ball travelling at $5 \, \text{m s}^{-1}$ strikes the only standing pin straight on. The pin flies backward at $7 \, \text{m s}^{-1}$. Calculate:
   (a) the velocity of the bowling ball after the collision
   (b) the loss of kinetic energy in this collision.
   (mass of bowling ball = 6.35 kg; mass of pin = 1 kg)

3. ▶ In a particle collision experiment, a mystery particle collides with a stationary neutron and sets the neutron into motion with a velocity of $1.5 \times 10^7 \, \text{m s}^{-1}$. The mystery particle arrived at a velocity of 1% of the speed of light, and recoiled after collision with a velocity of $7.5 \times 10^5 \, \text{m s}^{-1}$ in the opposite direction. Calculate the mass of the mystery particle, and identify it.

SUBJECT VOCABULARY

elastic collision a collision in which total kinetic energy is conserved

inelastic collision a collision in which total kinetic energy is not conserved

LEARNING OBJECTIVES

- Apply the conservation of linear momentum to situations in two dimensions.
- Analyse collisions in two dimensions.
- Calculate impulses and changes in momentum.

So far, we have only considered the conservation of linear momentum in one-dimensional collisions, where all objects move forwards and/or backwards along the *same* straight line. This is an unusual situation, and we need to be able to work with more complex movements. Helpfully, momentum is conserved in each dimension separately. So, we resolve vector movements entering a collision into components in each dimension and then calculate following the conservation of momentum in each dimension. After this, we can recombine component vectors to give us an overall vector after a collision. A real world example, as in **fig B**, will illustrate this best.

PRACTICAL SKILLS   CP9

Investigating impulse

motion sensor

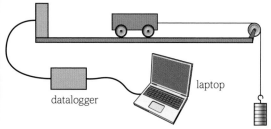

▲ **fig A** Measuring how impulse changes the momentum of a trolley

In **Book 1, Section 1C.1**, we saw how you can investigate the change in momentum over time for a trolley that is subject to a constant accelerating force. Using the same apparatus, you could again record how different forces acting over different time periods cause the trolley to accelerate to different velocities. From these results, you can calculate the impulse applied in each case. As $I = F\Delta t = \Delta p = \Delta mv$, a graph of impulse on the y-axis against change in velocity on the x-axis should give a straight best fit line through the origin. This straight line verifies the impulse equation, and the gradient of it will give the mass of the accelerating trolley and weights.

Safety Note: Put a 'catch box' full of soft material under the hanging masses to stop them falling on to feet and use an end stop to prevent the trolley falling off the runway.

EXAM HINT

Make sure you have a good understanding of this practical as your understanding of the experimental method may be assessed in your exams.

IMPULSE

The product of a force applied for a known time ($\boldsymbol{F} \times \Delta t$) is known as the **impulse**, and this is equal to the change in momentum:

impulse (Ns) = force (N) × time (s)
            = change in momentum (kg m s^{-1})

impulse   = $\boldsymbol{F} \times \Delta t = \Delta \boldsymbol{p}$

To stop something moving, we need to remove all of its momentum. This idea allows us to calculate the impulse needed to stop an object moving. If we know how long a force is applied, we could work out the size of that force.

WORKED EXAMPLE 1

What is the impulse needed to accelerate a 1000 kg car from rest to 25 m s^{-1}?

$$\boldsymbol{p} = m \times \boldsymbol{v}$$

At the start, the car is at rest, so has no momentum. Therefore the change in momentum will equal its final momentum:

$$\Delta \boldsymbol{p} = 1000 \times 25$$

$$\text{impulse} = \boldsymbol{F} \times \Delta t = \Delta \boldsymbol{p}$$

$$\boldsymbol{I} = 25\,000 \text{ kg m s}^{-1}$$

If the car needed to stop in 3.8 seconds, what force would the brakes need to apply?

At the end, the car is at rest, so has no momentum. Therefore the change in momentum will equal its initial momentum:

$$\boldsymbol{F}t = m\Delta \boldsymbol{v}$$

$$m\Delta \boldsymbol{v} = 25\,000 \text{ kg m s}^{-1}$$

$$\boldsymbol{F} = \frac{m\Delta \boldsymbol{v}}{t}$$

$$\boldsymbol{F} = \frac{25\,000}{3.8}$$

$$\boldsymbol{F} = 6600 \text{ N to 2 significant figures (sf)}$$

COLLISION VECTORS

Momentum is a vector quantity. As with all vectors, we can resolve momentum into a right-angle pair of vector components. And we can add components together in two dimensions to find an overall vector. A spacecraft is moving through empty space at 8 m s^{-1}. A meteoroid, travelling at 15 m s^{-1}, comes from behind and at an angle of 45° to the line of movement of the rocket, crashes into the rocket and becomes embedded in it. The rocket has a mass of 350 kg and the meteorite mass is 20 kg. We can calculate the velocity of the rocket (**fig B**) after the collision.

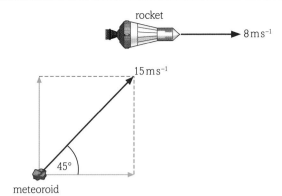

fig B A collision in two dimensions.

Before collision:

Parallel to rocket motion:

$$v_{meteorite} = 15 \cos 45° = 10.6 \text{ m s}^{-1}$$
$$p_{meteorite} = 20 \times 10.6 = 212 \text{ kg m s}^{-1}$$
$$p_{rocket} = 350 \times 8 = 2800 \text{ kg m s}^{-1}$$
$$p_{parallel} = 2800 + 212 = 3012 \text{ kg m s}^{-1}$$

Perpendicular to rocket motion:

$$v_{meteorite} = 15 \sin 45° = 10.6 \text{ m s}^{-1}$$
$$p_{meteorite} = 20 \times 10.6 = 212 \text{ kg m s}^{-1}$$
$$p_{rocket} = 350 \times 0 = 0 \text{ kg m s}^{-1}$$
$$p_{perpendicular} = 0 + 212 = 212 \text{ kg m s}^{-1}$$

After collision:

Vector sum of momenta (fig C):

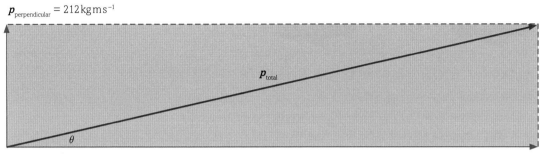

fig C Vector sum of total momentum in two dimensions.

$$p_{total} = \sqrt{(3012^2 + 212^2)} = 3019 \text{ kg m s}^{-1}$$

$$v_{after} = \frac{p_{total}}{(m_{rocket} + m_{meteorite})} = \frac{3019}{(350 + 20)} = 8.16 \text{ m s}^{-1}$$

Angle of momentum (i.e. direction of velocity) after collision:

$$\theta = \tan^{-1}\left(\frac{212}{3012}\right) = 4.0°$$

So, the spacecraft with embedded meteorite carries on at 8.16 m s^{-1} at an angle of 4.0° off the original direction of motion.

PRACTICAL SKILLS
CP10

Investigating 2D collisions

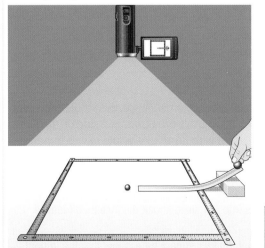

fig D Video analysis of collisions in 2D.

You can investigate two-dimensional collisions in the school laboratory. We saw in **Book 1, Section 1A.2** that by analysing video footage of an object's movement, frame by frame, we can calculate any changes in velocity. With measurement scales in two dimensions, the components of velocity in each dimension can be isolated. This means that separate calculations can be made in each dimension, in order to verify the conservation of momentum in 2D.

Safety Note: Use a heavy stand and a clamp to secure the camera so that it cannot fall over.

EXAM HINT

Collision and momentum exam questions often ask **Show that** ... In 'show that' questions, you must state the equations you use. Then substitute in values and calculate a final answer that rounds to the approximate value in the question. Give the answer to 1 significant figure more than given in the question to prove you have calculated it yourself, and that it matches with the number in the question.

For example, for the calculation on the left, an exam question could be 'Show that the total momentum after the collision is 3020 kg ms^{-1}.' Our calculations would show:

momentum = 3019 kg ms^{-1}

We should then conclude the answer with:

p_{total} = 3019 = 3020 kg ms^{-1} (3sf)

EXAM HINT

Make sure you have a good understanding of this practical as your understanding of the experimental method may be assessed in your exams.

▲ **fig E** The comet Tempel 1 was hit by NASA's Deep Impact probe.

DEEP SPACE COLLISION

On 4 July 2005, NASA's Deep Impact mission succeeded in crashing a spacecraft into a comet called Tempel 1 (**fig E**). For that mission, the impactor spacecraft had a mass of 370 kg compared with the comet's mass of 7.2×10^{13} kg, so there would have been an insignificant change in the comet's trajectory. Deep Impact was purely intended to study the comet's composition. However, there is an asteroid named Apophis which has a small chance of colliding with Earth in 2035, 2036, or maybe 2037, and there have been some calls for a mission to crash a spacecraft into Apophis in order to move it out of the crash line. The mass of this asteroid is 6.1×10^{10} kg and it is travelling at 12.6 km s^{-1}. It has been claimed that a collision by a 4000 kg impactor craft travelling at 6 km s^{-1} could change the path of this asteroid enough to ensure it would not hit Earth. If this impactor collided with Apophis at right angles, we can calculate the change in angle of the asteroid (**fig F**).

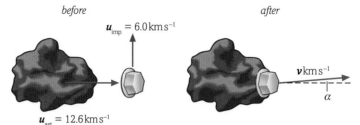

before $u_{imp} = 6.0$ km s^{-1} *after* v km s^{-1} α

$u_{ast} = 12.6$ km s^{-1}

▲ **fig F** Could we hit an asteroid hard enough to save Earth from Asteroid Impact Hazards?

Before collision:

$$p_{ast} = m_{ast}u_{ast} = 6.1 \times 10^{10} \times 12.6 \times 10^3 = 7.69 \times 10^{14} \text{ kg m s}^{-1}$$

$$p_{imp} = m_{imp}u_{imp} = 4 \times 10^3 \times 6 \times 10^3 = 2.4 \times 10^7 \text{ kg m s}^{-1}$$

total momentum p after impact α $p_{imp} = 2.4 \times 10^7$ kg m s^{-1}

$p_{ast} = 7.69 \times 10^{14}$ kg m s^{-1}

▲ **fig G** The vector sum of momentum components after asteroid impact.

The momentum of the combined object after the impactor embeds in the asteroid is the vector sum of the two initial momenta, which are at right angles to each other.

After collision:

$$p_{after} = \sqrt{(p_{ast}^2 + p_{imp}^2)} = \sqrt{((7.69 \times 10^{14})^2 + (2.4 \times 10^7)^2)}$$

$$= 7.69 \times 10^{14} \text{ kg m s}^{-1}$$

$$\therefore \quad v_{after} = \frac{p_{after}}{m_{total}} = \frac{7.69 \times 10^{14}}{(6.1 \times 10^{10} + 4 \times 10^3)}$$

$$= 12.6 \text{ km s}^{-1} \text{ (3 significant figures)}$$

There is no significant change in the magnitude of the asteroid's velocity. Is there a significant change in its direction?

Angle of momentum after:

$$\alpha = \tan^{-1}\left(\frac{2.4 \times 10^7}{7.69 \times 10^{14}}\right) = 1.79 \times 10^{-6} \text{ }^\circ$$

Although less than two microdegrees sounds like an insignificantly small angle, this would represent a change in position of nearly 30 km as Apophis crosses the Earth's orbit from one side of the Sun to the other. This might be just enough to prevent a collision with Earth that would have a hundred times more energy than all the explosives used in the Second World War.

DID YOU KNOW?

Archer fish catch insect prey by squirting water droplets into the air to knock the insects off leaves above the surface.

Calculations of the mass and velocity of the water droplet, and its impact time, show that the impact force can be ten times stronger than the insect's grip on the leaf.

CHECKPOINT

1. (a) What is the impulse needed to stop a car that has a momentum of 22 000 kg m s^{-1}?
    (b) If the car brakes could apply a force of 6800 N, how long would it take to bring the car to a stop?

2. ▶ In a pool shot, the cue ball has a mass of 0.17 kg. It travels at 6.00 m s^{-1} and hits the stationary black ball in the middle of one end of the table. The black ball, also of mass 0.17 kg, travels away at 45° and 4.24 m s^{-1}, ending up in the corner pocket.

    SKILLS ▶ CRITICAL THINKING

    (a) By resolving the components of the black ball's momentum, find out what happens to the cue ball.
    (b) Is this an elastic or inelastic collision?

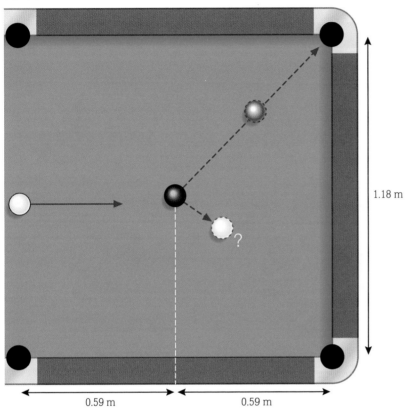

▲ **fig H** 2D momentum calculations can be very important in winning at pool.

3. Calculate how fast the impactor spacecraft in **fig F** would have to be travelling if it is to alter the Apophis asteroid's trajectory by one degree. Comment on the answer.

4. ▶ For the experimental set up shown in **fig D**, suggest two improvements that could be made in order to improve the accuracy of the resolved vectors that would be observable on the video stills.

    SKILLS ▶ INNOVATION

SUBJECT VOCABULARY

impulse force acting for a certain time causing a change in an object's momentum

    impulse = $F \times \Delta t$

ASTEROIDS

SKILLS CRITICAL THINKING, PROBLEM SOLVING, ANALYSIS, INTERPRETATION, ADAPTIVE LEARNING

This poster from the Planetary Society explains the preparations needed for an asteroid impact with the Earth.

PUBLIC INFORMATION POSTER

DEFENDING EARTH

With advanced planning and preparation, we could prevent a disastrous impact from an asteroid or comet. The Planetary Society breaks it down into these five steps for saving the world.

2. Track

If we find a near-Earth object, how do we know if it will hit Earth? We need to map its orbit by taking repeated observations. A number of missions, observatories, and systems track the orbits of NEOs, and more are in development.

1. Find

Astronomers use ground- and space-based telescopes to spot NEOs and have found 90% of the largest ones. Infrared imaging also helps find objects that are too dark to see from their reflected light.

THE PLANETARY SOCIETY
Find out more at
planetary.org/defense

3. Characterize

By characterizing the spin rate, composition, and physical properties of potentially hazardous NEOs, we can better know how to deflect them. Awardees of The Planetary Society's Shoemaker NEO Grant Program are making tremendous contributions in this area.

28
Ni
Nickel

14
Si
Silicon

4. Deflect

There is a variety of possible techniques for deflecting a potential impact, but all need more development and testing:

Slow gravity tractor: A massive spacecraft follows next to the near-Earth object and uses the spacecraft's gravity to pull the object off its collision course.

Kinetic impactor: A swarm of spacecraft slam into the object to knock it off course.

Laser ablation: A spacecraft uses lasers to vaporize rock on the object, creating jets that push it off course. The Planetary Society is researching this technique with the University of Strathclyde through their Laser Bees project.

5. Coordinate and Educate

An asteroid impact is a worldwide issue that requires immense advance coordination and education. The Planetary Society is taking an active role by working with governments around the world, hosting conferences, doing public outreach, and supporting volunteer efforts.

What about the nuclear option? Detonating nuclear devices on or beside an asteroid may be the only viable technique we have today for deflecting an asteroid. But this comes with challenges, including political opposition and the danger of fragment impacts.

From the Planetary Society *http://www.planetary.org/explore/projects/planetary-defense/*

SCIENCE COMMUNICATION

The poster was produced by the Planetary Society. It aims to explain the issues connected with a possible asteroid collision with the Earth.

1 (a) Discuss the tone and level of vocabulary and level of scientific detail in the poster. Who is the intended audience?

    (b) Discuss which of the images are the most useful to support the text, and which do not add so much.

2 Explain which of the five sections on the poster explains the most scientific ideas. Why do you think this section has the most scientific ideas?

INTERPRETATION NOTE

Consider which sections of the poster describe science that has been tested the most. Which sections are well understood? Which need more research?

PHYSICS IN DETAIL

Now we will look at the physics in detail. Some of these questions will link to topics elsewhere in this book, so you may need to combine concepts from different areas of physics to work out the answers.

3 (a) Look at section *1. Find* in the poster. What is a NEO?

    (b) Explain two of the difficulties in finding NEOs.

4 Consider an asteroid 1000 km in diameter, with a structure of iron and rock. The overall density of such asteroids is about $2\,000\,\text{kg}\,\text{m}^{-3}$.

    (a) Estimate the volume of the asteroid. Why is your answer an estimate and not the exact answer?

    (b) Calculate an estimate for its mass.

    (c) Imagine the asteroid travelled directly towards the Earth at $10\,000\,\text{m}\,\text{s}^{-1}$ and collided and embedded into the surface of the Earth. Calculate the change in speed of the Earth. The mass of the Earth is $5.97 \times 10^{24}\,\text{kg}$.

5 A Planetary Society scientist suggests we try to blow the asteroid apart with a nuclear explosion (as in *4. Deflect*) before it hits the Earth. If the bomb can be set off 24 hours before collision with the Earth, and can split the asteroid into two equal pieces, calculate the force that the nuclear explosion, lasting for 2 seconds, would need to apply to send the two parts off course enough to save the Earth. The Earth's radius is 6400 km.

PHYSICS TIP

Think about what would determine the motion of the broken pieces after the explosion. Consider conservation of momentum.

ACTIVITY

Write a short talk for a member of the Planetary Society to give to a school age audience explaining the physics of the 'kinetic impactor' deflection method. This talk could be a part of the activities in *5. Coordinate and Educate*.

THINKING BIGGER TIP

You may need to do some further research about the 'kinetic impactor' idea. Concentrate on the conservation of momentum and the vector additions involved.

[Note: In questions marked with an asterisk (*), marks will be awarded for your ability to structure your answer logically, showing how the points that you make are related or follow on from each other.]

1 An inelastic collision:

  A conserves momentum but not kinetic energy

  B conserves momentum and kinetic energy

  C need not conserve energy

  D need not conserve momentum. [1]

(Total for Question 1 = 1 mark)

2 A tennis ball travelling with the momentum of $4.2\,\text{kg m s}^{-1}$ is hit by a tennis racquet. The force of $56\,\text{N}$ from the racquet causes the tennis ball to travel back in the opposite direction with the momentum of $5.8\,\text{kg m s}^{-1}$. How long is the ball in contact with the racquet?

  A $0.029\,\text{s}$

  B $0.10\,\text{s}$

  C $0.18\,\text{s}$

  D $5.6\,\text{s}$ [1]

(Total for Question 2 = 1 mark)

3 In order to calculate the kinetic energy of a non-relativistic particle, we would need to know its:

  A mass only

  B mass and momentum

  C acceleration and momentum

  D velocity and acceleration. [1]

(Total for Question 3 = 1 mark)

4 In an experiment to accelerate a trolley along a runway, a light gate measured the time the trolley took to pass through it. There were five repeats of the same acceleration test and it took the following readings:

    $0.87\,\text{s}, 0.89\,\text{s}, 0.65\,\text{s}, 0.76\,\text{s}, 0.77\,\text{s}$

Which of these is the correct percentage error for this set of time readings?

  A 0.12%

  B 0.15%

  C 15%

  D 30% [1]

(Total for Question 4 = 1 mark)

5 A spacecraft called Deep Space 1, mass $486\,\text{kg}$, uses an 'ion-drive' engine. This type of engine is designed to be used in deep space.

The following statement appeared in a website.

> The ion propulsion system on Deep Space 1 expels $0.13\,\text{kg}$ of xenon propellant each day. The xenon ions are expelled from the spacecraft at a speed of $30\,\text{km s}^{-1}$. The speed of the spacecraft is predicted to initially increase by about $8\,\text{m s}^{-1}$ each day.

Use a calculation to comment on the prediction made in this statement. [4]

(Total for Question 5 = 4 marks)

6 (a) Explain what is meant by the principle of conservation of momentum. [2]

  (b) The picture shows a toy car initially at rest with a piece of modelling clay attached to it.

  A student carries out an experiment to find the speed of a pellet fired from an air rifle. The pellet is fired horizontally into the modelling clay. The pellet remains in the modelling clay as the car moves forward. The motion of the car is filmed for analysis.

  The car travels a distance of $69\,\text{cm}$ before coming to rest after a time of $1.3\,\text{s}$.

  (i) Show that the speed of the car immediately after being struck by the pellet was about $1\,\text{m s}^{-1}$. [2]

  (ii) State an assumption you made in order to apply the equation you used. [1]

  (iii) Show that the speed of the pellet just before it collides with the car is about $120\,\text{m s}^{-1}$

    mass of car and modelling clay = $97.31\,\text{g}$

    mass of pellet = $0.84\,\text{g}$ [3]

  (c) The modelling clay is removed and is replaced by a metal plate of the same mass. The metal plate is fixed to the back of the car. The experiment is repeated but this time the pellet bounces backwards.

  *(i) Explain why the speed of the toy car will now be greater than in the original experiment. [3]

  (ii) The film of this experiment shows that the pellet bounces back at an angle of $72°$ to the horizontal. Explain why the car would move even faster if the pellet bounced directly backwards at the same speed. [1]

(d) The student tests the result of the first experiment by firing a pellet into a pendulum with a bob made of modelling clay. They calculate the energy transferred.

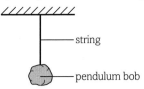

The student's data and calculations are shown:

> *Data:*
> *mass of pellet = 0.84 g*
> *mass of pendulum and pellet = 71.6 g*
> *change in vertical height of pendulum = 22.6 cm*
>
> *Calculations:*
> *change in gravitational potential energy of pendulum and pellet*
> $= 71.6 \times 10^{-3}\,\text{kg} \times 9.81\,\text{N kg}^{-1} \times 0.226\,\text{m} = 0.16\,\text{J}$
> *therefore kinetic energy of pendulum and pellet immediately after collision = 0.16 J*
> *therefore kinetic energy of pellet immediately before collision = 0.16 J*
> *therefore speed of pellet before collision = 19.5 m s^{-1}*

There are no mathematical errors but the student's answer for the speed is too small.

Explain which of the statements in the calculations are correct and which are not.   [4]

(Total for Question 6 = 16 marks)

7 James Chadwick is credited with 'discovering' the neutron in 1932.

Beryllium was bombarded with alpha particles, knocking neutrons out of the beryllium atoms. Chadwick placed various targets between the beryllium and a detector. Hydrogen and nitrogen atoms were knocked out of the targets by the neutrons and the kinetic energies of these atoms were measured by the detector.

(a) The maximum energy of a nitrogen atom was found to be 1.2 MeV.

Show that the maximum velocity of the atom is about $4 \times 10^6\,\text{m s}^{-1}$.

mass of nitrogen atom = $14u$,
where $u = 1.66 \times 10^{-27}\,\text{kg}$ [3]

(b) The mass of a neutron is Nu (where N is the relative mass of the neutron) and its initial velocity is $\boldsymbol{x}$. The nitrogen atom, mass $14u$, is initially stationary and is then knocked out of the target with a velocity, $\boldsymbol{y}$, by a collision with a neutron.

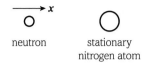

neutron   stationary nitrogen atom

(i) Show that the velocity, $\boldsymbol{z}$, of the neutron after the collision can be written as
$$z = \frac{N\boldsymbol{x} - 14\boldsymbol{y}}{N}$$ [3]

(ii) The collision between this neutron and the nitrogen atom is elastic. What is meant by an elastic collision?   [1]

(iii) Explain why the kinetic energy E_k of the nitrogen atom is given by
$$E_k = \frac{Nu(\boldsymbol{x}^2 - \boldsymbol{z}^2)}{2}$$ [2]

(c) The two equations in (b) can be combined and $\boldsymbol{z}$ can be eliminated to give
$$y = \frac{2N\boldsymbol{x}}{N + 14}$$

(i) The maximum velocity of hydrogen atoms displaced by neutrons in the same experiment was $3.0 \times 10^7\,\text{m s}^{-1}$. The mass of a hydrogen atom is $1u$. Show that the relative mass N of the neutron is 1.   [3]

(ii) This equation cannot be applied to all collisions in this experiment. Explain why.   [1]

(Total for Question 7 = 13 marks)

TOPIC 5 FURTHER MECHANICS

CHAPTER 5B CIRCULAR MOTION

What is it that makes the swing carousel so much fun? How do the engineers who build it have confidence that it will operate safely? The motion of objects in circles is very common in everyday life. We see the Moon orbiting the Earth, and the Earth orbiting the Sun, whilst the Earth itself is making us rotate in a circle every day. The mathematics of the forces causing this motion is simple, but applying this to real-life situations, like the amount of tension in the chains of a swing carousel, can often leave people feeling confused.

In this chapter, we will learn how simple trigonometry and basic mechanics can generate the mathematics that we need to be able to analyse and predict the circular motions of many things. This includes things from subatomic particles being accelerated in the Large Hadron Collider, to the routing of a spacecraft travelling to Mars. Indeed, very simple circular motion calculations on entire galaxies led scientists to the idea that there must be dark matter throughout the Universe.

MATHS SKILLS FOR THIS CHAPTER

- **Use of trigonometric functions** (*e.g. in deriving the equations for centripetal acceleration*)
- **Use of an appropriate number of significant figures** (*e.g. calculating angular velocities of planets*)
- **Use of small angle approximations** (*e.g. in deriving the equations for centripetal acceleration*)
- **Changing the subject of an equation** (*e.g. finding the time period of orbits from a known angular velocity*)
- **Translating information between numerical and graphical forms** (*e.g. finding the square of the angular velocity to plot a graph*)
- **Determining the slope of a linear graph** (*e.g. finding the gradient of the line in a centripetal force experiment analysis*)

What prior knowledge do I need?

- The radian
- Tension forces

Topic 1A (Book 1: IAS)

- Speed and acceleration
- How to add vectors
- How to resolve vectors
- Newton's laws of motion

What will I study in this chapter?

- How to use both radians and degrees in angle measurements
- The concept of angular velocity, and how to calculate it
- Centripetal acceleration and how to derive its equations
- The need for a centripetal force to enable circular motion
- Calculations of centripetal force

What will I study later?

Topic 7B

- The circular motion of charged particles in magnetic fields
- Particle accelerators
- Particle detectors
- The Large Hadron Collider

Topic 10A

- Simple harmonic motion

Topic 11A

- Gravitational fields and orbital motion

In our study of wave phase in **Book 1, Section 3B.1**, we learned how angles can be measured in radians as well as in degrees, and how to convert between these two units. In this section, we will see how we can extend that to measure movements around a circle.

GOING ROUND IN CIRCLES

For an object moving in a circle, we often need to measure where it is around that circle. For example, to consider the relative positions of planets in their orbits at a particular time, we need to be able to state where each one is (ignoring, in this example, the fact that planetary orbits are not perfect circles!).

Angles measured in degrees are used extensively in navigation to locate places. We also use them to describe the difference between moving from one starting point to two possible destinations. This is measuring angular displacement on the surface of the Earth. Each degree is subdivided into 60 'arcminutes' and each of those arcminutes into 60 'arcseconds'.

▲ **fig A** Measuring angles in degrees.

When we are measuring rotation, we often use the alternative unit to measure angles – the **radian**. The circle itself defines this. Imagine an object moves around part of the circumference of a circle. The angle through which it moves, measured in radians, is defined as the distance it travels, divided by its distance from the centre of the circle (the radius). If the radius of the circle were one metre, then the distance the object travels around the circumference (also in metres) would be equal to the angle swept out in radians.

$$\text{angle (in radians)} = \frac{\text{length of arc}}{\text{radius of arc}}$$

$$\theta = \frac{s}{r}$$

So, for a complete circle, in which the circumference is equal to $2\pi r$, the angle swept out would be:

$$\theta = \frac{2\pi r}{r} = 2\pi \text{ radians}$$

This means that the angle will be 1 radian (rad) if the distance moved around the circle is the same as the radius; just over $\frac{1}{6}$ of the distance around the circumference.

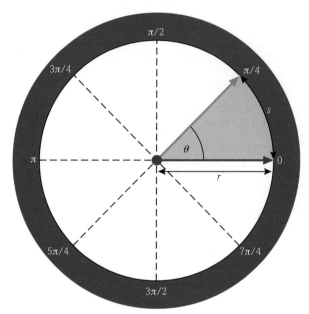

▲ **fig B** Measuring angles in radians.

Angular displacement is the vector measurement of the angle through which something has moved. The standard convention is that anticlockwise rotation is a positive number and clockwise rotation is a negative number.

ANGLE IN RADIANS	ANGLE IN DEGREES
0	0
$\pi/4$	45
$\pi/2$	90
$3\pi/4$	135
π	180
$5\pi/4$	225
$3\pi/2$	270
$7\pi/4$	315
2π	360

table A Angles measured in degrees and radians.

ANGULAR VELOCITY

An object moving in a circle sweeps out a particular angle in a particular time, depending upon how fast it is moving. The rate at which the angular displacement changes is called the **angular velocity, ω**. So, angular velocity is measured in rad s^{-1}, and is defined mathematically by:

$$\omega = \frac{\Delta\theta}{\Delta t}$$

If the object completes a full circle (2π radians) in a time period, T, then the angular velocity is given by:

$$\omega = \frac{2\pi}{T}$$

$$\therefore \quad T = \frac{2\pi}{\omega}$$

The frequency of rotation is the reciprocal of the time period.

$$f = \frac{1}{T}$$

$$\therefore \quad \omega = 2\pi f$$

INSTANTANEOUS VELOCITY

Instead of thinking about the angular movement, let us consider the actual velocity of the moving object (sometimes called the 'instantaneous velocity'). We know that $v = \frac{s}{t}$ and from the definition of the angle in radians $\theta = \frac{s}{r}$, so that $s = r\theta$.

Thus: $\quad v = \frac{r\theta}{t}$

$$v = r\omega$$

WORKED EXAMPLE 1

In **fig D** we can see a geostationary satellite orbiting the Earth. What is its angular velocity?

To find the angular velocity, remember that it completes an orbit at the same rate as the Earth revolves, so one full circle every 24 hours.

$$\omega = \frac{2\pi}{T}$$

$$\omega = \frac{2\pi}{(24 \times 60 \times 60)} = \frac{2\pi}{86\,400}$$

$$\omega = 7.27 \times 10^{-5}\,\text{rad s}^{-1}$$

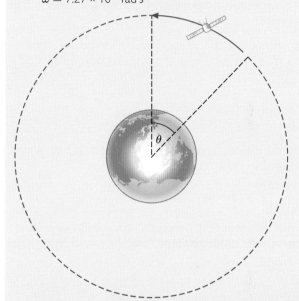

▲ **fig D** How quickly does a satellite rotate through a certain angle?

If the radius of the Earth is 6400 km and the satellite in **fig D** is in orbit 35 600 km above the Earth's surface, what is the velocity of the satellite?

From before, $\omega = 7.27 \times 10^{-5}\,\text{rad s}^{-1}$

$$v = r\omega = (6400 + 35\,600) \times 10^{3} \times 7.27 \times 10^{-5}$$

$$\therefore \qquad v = 3050\,\text{m s}^{-1}$$

CENTRIPETAL ACCELERATION

Velocity is a vector, and so it is correctly described by both its magnitude and direction. An acceleration can change either of these, or both. An object moving in a circle may travel at a constant speed (and a constant angular velocity) but the direction it is moving in must constantly change. This means it is constantly accelerating. As this acceleration represents the changes in direction around the circle, it is called the **centripetal acceleration, a**. To calculate the centripetal acceleration, we must consider how quickly the direction, and therefore the velocity, is changing.

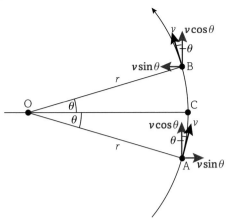

▲ **fig E** Vector components of velocity lead us to the centripetal acceleration equation.

At the arbitrary positions of the rotating object, A, and at a time t later, B, we consider the components of the object's velocity, in the x and y directions.

As A and B are equal distances above and below point C, the vertical velocity component, v_y, is the same in both cases:

$$v_y = v \cos \theta$$

So the vertical acceleration is zero:

$$a_y = 0$$

Horizontally, the magnitude of the velocity is equal at both points, but in opposite directions:

at A: $v_x = v \sin \theta$

at B: $v_x = -v \sin \theta$

So, the acceleration is just the horizontal acceleration, calculated as change in velocity divided by time:

$$a_x = \frac{2v \sin \theta}{t}$$

From the definition of angular velocity above:

$$v = \frac{r\theta}{t}$$

$$\therefore \quad t = \frac{r\theta}{v}$$

Here, the angle moved in time t is labelled as 2θ, so:

$$t = \frac{r2\theta}{v}$$

$$\therefore \quad a_x = \frac{v2v \sin \theta}{r2\theta} = \frac{v^2}{r} \frac{\sin \theta}{\theta}$$

This must be true for all values of θ, and as we want to find the instantaneous acceleration at any point on the circumference, we must consider the general answer as we reduce θ to zero. In the limit, as θ tends to zero:

$$\frac{\sin \theta}{\theta} = 1$$

$\therefore$ centripetal acceleration, $a = \dfrac{v^2}{r}$

From the definition of the instantaneous velocity above:

$$v = r\omega$$

$$\therefore \quad a = \frac{(r\omega)^2}{r}$$

$\therefore$ centripetal acceleration, $a = r\omega^2$

The centripetal acceleration in this case is just the horizontal acceleration, as we considered the object in a position along a horizontal radius. Following a similar derivation at any point around the circle will always have identical components of velocity that are perpendicular to the radius on either side of the point being considered. Thus, the centripetal acceleration is always directed towards the centre of the circle.

WORKED EXAMPLE 2

What is the centripetal acceleration of the satellite in **fig D**?

$$a = \frac{v^2}{r} = \frac{(3050)^2}{(6400 + 35\,600) \times 10^3}$$

$$a = 0.22 \, \mathrm{m\,s^{-2}}$$

or:

$$a = r\omega^2 = (6400 + 35\,600) \times 10^3 \times (7.27 \times 10^{-5})^2$$

$$a = 0.22 \, \mathrm{m\,s^{-2}}$$

CHECKPOINT

SKILLS ▷ ANALYSIS

1. ▷ Convert:
   (a) 4π radians into degrees
   (b) 36° into radians.

2. What is the angular velocity of an athletics hammer if the athlete spins it at a rate of three revolutions per second?

3. Old vinyl records were played at one of three speeds. Calculate the angular velocity of each:
   (a) (i)   33 revolutions per minute
       (ii) 45 revolutions per minute
       (iii) 78 revolutions per minute.
   (b) Vinyl records played at the speeds in part (a) (i) were usually 12 inches (or 30 cm) in diameter. What would be the centripetal acceleration of a point on the outside circumference of a record such as this?

4. A man standing on the Equator will be moving due to the rotation of the Earth.
   (a) What is his angular velocity?
   (b) What is his instantaneous velocity?
   (c) What is his centripetal acceleration?

5. ▷ What is the percentage error in the measurement system that uses 6400 mils for a complete circle?

SUBJECT VOCABULARY

radian a unit of angle measurement, one radian is equivalent to 57.3 degrees

angular displacement the vector measurement of the angle through which something has moved

angular velocity, ω the rate at which the angular displacement changes; unit, radians per second

centripetal acceleration, a the acceleration towards the centre of a circle that corresponds to the changes in direction to maintain an object's motion around that circle

LEARNING OBJECTIVES

- Explain that a centripetal force is required to produce and maintain circular motion.
- Use the equations for centripetal force.

In **Section 5B.1**, we saw how an object moving in a circle must be constantly accelerating towards the centre of the circle in order to maintain its motion around the circle.

WHY CIRCULAR MOTION?

When a hammer thrower moves an athletics hammer around in a circle, the hammer has an angular velocity. When the thrower lets the hammer go, it will fly off, following the straight line in the direction that it was moving at the instant of release. This direction is always along the edge of the circle (a tangent) at the point when it was released.

▲ **fig A** The instantaneous velocity of an object moving in a circle is tangential to the circle. When there is no resultant force, velocity will be constant, so it moves in a straight line, as this hammer will do when the athlete lets go.

The hammer is moved in a circle at a constant speed and so the magnitude of the velocity is always the same. However, the direction of the velocity is constantly changing. This means that the vector of velocity is constantly changing, and a change in velocity is an acceleration. Newton's first law tells us that acceleration will only happen if there is a resultant force. The hammer is constantly being pulled towards the centre of the circle. In this example, the force providing this pull is the tension in the string (or chain). For any object moving in a circle, there must be a resultant force to cause this acceleration, and it is called the **centripetal force**.

CENTRIPETAL FORCE

The mathematical formula for the centripetal force on an object moving in a circle can be found from Newton's second law, and the equation we already have for the centripetal acceleration:

$$F = ma \quad \text{and} \quad a = \frac{v^2}{r}$$

$$F = \frac{mv^2}{r}$$

$$\text{centripetal force} = \frac{\text{mass} \times (\text{velocity})^2}{\text{radius}}$$

EXAM HINT

A centrifuge is a machine for spinning things. There is no 'centrifugal' force - do not use the phrase 'centrifugal force'.

Without a centripetal force acting, Newton's 1st law will make an object continue in a straight line. This can confuse people into thinking that a force is pushing an object outwards. In fact, this shows a lack of force; if a force were present, this could keep it moving in a circle.

▲ **fig B** Astronauts are subject to extreme acceleration forces. These forces are simulated in training by the centripetal force in a giant centrifuge.

Noting that $v = r\omega$, there is an alternative equation for centripetal force in terms of angular velocity:

$$F = \frac{mv^2}{r} = \frac{m(r\omega)^2}{r}$$

$$F = mr\omega^2$$

The resultant centripetal force needed will be larger if:

- the rotating object has more mass
- the object rotates faster
- the object is closer to the centre of the circle.

WORKED EXAMPLE 1

Estimate the centripetal force on an astronaut in the astronaut training centrifuge (see **fig B**) if his capsule rotated once every 2 seconds.

Estimate of radius of revolution: $r = 6$ m.

Estimate of astronaut's mass: $m = 80$ kg.

$$\text{velocity, } v = \frac{s}{t} = \frac{2\pi(6)}{2} = 18.8 \text{ m s}^{-1}$$

$$\text{centripetal force, } F = \frac{mv^2}{r} = \frac{(80) \times (18.8)^2}{6} = 4700 \text{ N}$$

This is about 6 times his weight.

If the operator of the centrifuge were to increase its rate of rotation to once every second, what would the astronaut's angular velocity, centripetal force and acceleration now be?

$$\text{angular velocity, } \omega = 2\pi f = 6.28 \text{ rad s}^{-1}$$

$$\text{centripetal force, } F = mr\omega^2 = (80) \times (6) \times (6.28)^2 = 18\,900 \text{ N}$$

$$\text{centripetal acceleration, } a = r\omega^2 = (6) \times (6.28)^2 = 237 \text{ m s}^{-2}$$

This is about 24 times the acceleration due to gravity and would probably kill the astronaut after a few seconds.

LEARNING TIP

Here is a summary of the equations for circular motion:

angular displacement, θ	$\theta = \dfrac{s}{r}$
angular velocity, ω	$\omega = 2\pi f = \dfrac{2\pi}{T} = \dfrac{v}{r}$
centripetal acceleration, a	$a = r\omega^2 = \dfrac{v^2}{r}$
centripetal force, F	$F = \dfrac{mv^2}{r}$
	$F = mr\omega^2$

PRACTICAL SKILLS

Investigating centripetal force

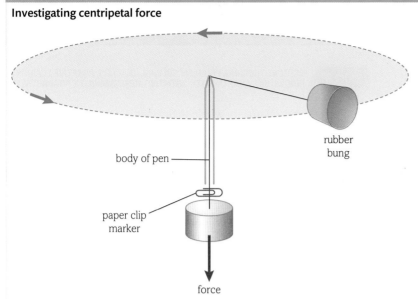

body of pen

rubber bung

paper clip marker

force

▲ **fig C** Experimental detail for centripetal force experiment.

You can investigate the centripetal force equation by spinning a rubber bung on a string around in a circle. The tension in the string, which is the centripetal force, will be provided by the hanging masses at the bottom of the vertical string and will be known. Spin the rubber bung around in a circle at a speed that keeps a paperclip marker in a constant position near the handle. The paperclip marker allows you to maintain a fixed length of string (radius) which you can measure. You will also need to measure the mass of the rubber bung. Your partner can then time ten revolutions in order to give you the angular velocity. Take angular velocity measurements for different forces (different numbers of hanging masses).

$$F = mr\omega^2$$
$$\therefore \quad \omega^2 = \frac{F}{mr}$$

A graph of ω^2 plotted against F should give a straight best-fit line.

The gradient of this line will be $\dfrac{1}{mr}$

> Safety Note: Wear eye protection and only use rubber bungs for whirling. Work away from windows and walls, preferably outdoors or in a large clear space.

CHECKPOINT

1. ▶ A roller coaster has a complete (circular) loop with a radius of 20 m. A 65 kg woman rides the roller coaster and the roller coaster car travels once round the loop in 4.5 seconds. What centripetal force does the woman experience?

2. A man with a mass of 75 kg standing on the Equator is moving because of the rotation of the Earth.
   (a) What centripetal force is required to keep him moving in this circle?
   (b) How does this compare with his weight?
   (c) How would the reaction force with the ground be different if he went to the South Pole? (Assume the Earth is a perfect sphere.)

SKILLS ▶ PROBLEM SOLVING

SUBJECT VOCABULARY

centripetal force the resultant force towards the centre of the circle to maintain an object's circular motion

ARTIFICIAL GRAVITY

SKILLS CRITICAL THINKING, PROBLEM SOLVING, ANALYSIS, ADAPTIVE LEARNING, INNOVATION, ADAPTABILITY, INITIATIVE, RESPONSIBILITY, PRODUCTIVITY

Astronauts on long-term space missions would benefit from the generation of artificial gravity within their ships. Common designs for this involve rotating the spaceship, but this movement can lead to motion sickness.

GRAVITY CALCULATOR WEBSITE

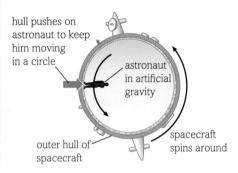

hull pushes on astronaut to keep him moving in a circle

astronaut in artificial gravity

spacecraft spins around

outer hull of spacecraft

▲ **fig A** This is how artificial gravity is generated in a spacecraft.

Artificial gravity, as it is usually conceived, is the inertial reaction to the centripetal acceleration that acts on a body in circular motion. Artificial-gravity environments are often characterized in terms of four parameters:

- **radius** from the centre of rotation (in metres)
- **angular velocity** or 'spin rate' (usually quoted in revolutions per minute)
- **tangential velocity** or 'rim speed' (in metres per second)
- **centripetal acceleration** or 'gravity level' (in multiples of the Earth's surface gravity).

Comfort criteria

Deliberate architectural design for the unusual conditions of artificial gravity ought to aid adaptation and improve the habitability of the environment (Hall).

Here is a summary of five research articles which consider the parameters leading to rotational discomfort:

AUTHOR	YEAR	RADIUS [M]	ANGULAR VELOCITY [RPM]	TANGENTIAL VELOCITY [M/S]	CENTRIPETAL ACCELERATION [G]	
		MIN.	MAX.	MIN.	MIN.	MAX.
Hill & Schnitzer	1962	not known	4	6	0.035	1.0
Gilruth	1969	12	6	not known	0.3	0.9
Gordon & Gervais	1969	12	6	7	0.2	1.0
Stone	1973	4	6	10	0.2	1.0
Cramer	1985	not known	3	7	0.1	1.0

Radius Because centripetal acceleration – the nominal artificial gravity – is directly proportional to radius, inhabitants will experience a head-to-foot 'gravity gradient'. To minimize the gradient, maximize the radius.

Angular velocity The cross-coupling of normal head rotations with the habitat rotation can lead to dizziness and motion sickness. To minimize this cross-coupling, minimize the habitat's angular velocity.

Graybiel conducted a series of experiments in a 15-foot-diameter 'slow rotation room' and observed:

In brief, at 1.0 rpm even highly susceptible subjects were symptom-free, or nearly so. At 3.0 rpm subjects experienced symptoms but were not significantly handicapped. At 5.4 rpm, only subjects with low susceptibility performed well and by the second day were almost free from symptoms. At 10 rpm, however, adaptation presented a challenging but interesting problem. Even pilots without a history of air sickness did not fully adapt in a period of twelve days.

On the other hand, Lackner and DiZio found that:

Sensory-motor adaptation to 10 rpm can be achieved relatively easily and quickly if subjects make the same movement repeatedly. This repetition allows the nervous system to gauge how the Coriolis forces generated by movements in a rotating reference frame are deflecting movement paths and endpoints, and to institute corrective adaptations.

Tangential velocity When people or objects move within a rotating habitat, they're subjected to Coriolis accelerations that distort the apparent gravity. For relative motion in the plane of rotation, the ratio of Coriolis to centripetal acceleration is twice the ratio of the relative velocity to the habitat's tangential velocity. To minimize this ratio, maximize the habitat's tangential velocity.

Centripetal acceleration The centripetal acceleration must have some minimum value to offer any practical advantage over weightlessness. One common criterion is to provide adequate floor traction. The minimum required to preserve health remains unknown.

From a webpage to calculate artificial gravity, maintained by Ted Hall: Hall, Theodore W. (2012). '*SpinCalc: An Artificial-Gravity Calculator in JavaScript*'; *www.artificial-gravity.com/sw/SpinCalc/SpinCalc.htm* (as at 24 October 2014).

23

SCIENCE COMMUNICATION

The extract above consists of information from an American space architect.

1. (a) Discuss the tone and level of vocabulary included in the article. Who is the intended audience?

   (b) Discuss the level of scientific detail included in the article, particularly considering the intended audience.

   (c) Compare the webpage of the extract with this more recent website http://spacearchitect.org/ considering style and presentation, in addition to content.

INTERPRETATION NOTE

The main basis for the website is a calculator that calculates the artificial gravity for various different sizes and speeds of rotation. You may find it helpful to visit the website to view it and the list of references it provides.

PHYSICS IN DETAIL

Now we will look at the physics in detail. Some of these questions will link to topics elsewhere in this book, so you may need to combine concepts from different areas of physics to work out the answers.

2. (a) Explain the author's suggestion that 'artificial gravity ... is proportional to radius'.

   (b) What would a 'gravity gradient' be? Explain how it would come about and why it would be minimised by maximising the radius.

3. (a) Convert the angular velocity in Gilruth's data into SI units.

   (b) Calculate the centripetal force on an 82 kg astronaut using Gilruth's data.

   (c) If this astronaut tried to walk by providing a tangential force of 350 N, calculate the moment caused by his foot, acting on the floor about the spacecraft's central axis.

4. Do some research on the Coriolis force effect, and use your research to explain the final sentence in the section on angular velocity.

5. Explain how a spaceship, in space, could be made to start rotating.

PHYSICS TIP

When you use the equations for centripetal acceleration, remember that the tangential velocity is dependent on radius.

ACTIVITY

Imagine that Mr Hall, the space architect, has been invited to give a talk at your school about artificial gravity. You have been asked to prepare some demonstrations to illustrate how the data in the experiments referred to above could have been measured. Prepare instructions for two demonstrations.

PHYSICS TIP

You should consider how to highlight where and how you will take measurements.

[Note: In questions marked with an asterisk (*), marks will be awarded for your ability to structure your answer logically, showing how the points that you make are related or follow on from each other.]

1 Which of the following is **not** a correct unit for angular velocity?

  A rad min^{-1}

  B degrees per minute

  C rad s

  D ° s^{-1} [1]

  (Total for Question 1 = 1 mark)

2 Considering the centripetal force to make a car drive around a bend, it is more likely to skid outwards if:

  A it has fewer passengers

  B it has worn tyres

  C it travels more slowly

  D it drives further from the inside of the curve. [1]

  (Total for Question 2 = 1 mark)

3 A particle moves in a circle, completing 14.5 complete revolutions in one minute. What is its angular velocity?

  A 0.24 rad s^{-1}

  B 1.52 rad s^{-1}

  C 14.5 rad s^{-1}

  D 87 rad s^{-1} [1]

  (Total for Question 3 = 1 mark)

4 A student did an experiment about the force on an object moving in a circle. The student spun a mass on a string in the air above his head so that the mass moved in a horizontal circle.

  Which force caused the mass to keep moving around in a circle?

  A centrifugal

  B gravity

  C tension

  D weight [1]

  (Total for Question 4 = 1 mark)

5 Kingda Ka was the highest roller coaster in the world in 2007. A train is initially propelled along a horizontal track by a hydraulic system. It reaches a speed of 57 m s^{-1} from rest in 3.5 s. It then climbs a vertical tower before falling back towards the ground.

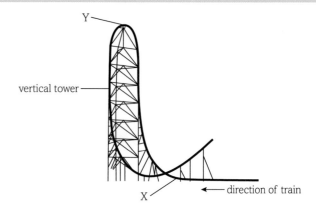

(a) Calculate the average force used to accelerate a fully loaded train along the horizontal track.
  Total mass of fully loaded train = 12 000 kg [2]

(b) Point X is just before the train leaves the horizontal track and moves into the first bend. Complete the free-body diagram below to show the two forces acting on a rider in the train at this point. [3]

(c) The mass of the rider is *m* and *g* is the acceleration of free fall. Just after point X, the reaction force of the train on the rider is 4 *mg* and can be assumed to be vertical. This is referred to as a *g*-force of 4 *g*.
  Show that the radius of curvature of the track at this point is about 100 m. [3]

(d) Show that the speed of the train as it reaches the top of the vertical tower is about 20 m s^{-1}. Assume that resistance forces are negligible.
  The height of the vertical tower is 139 m. [2]

(e) Riders will feel momentarily weightless if the vertical reaction force becomes zero.
  The track is designed so that this happens at point Y.
  Calculate the radius of the track at point Y. [2]

  (Total for Question 5 = 12 marks)

6 Astronauts can be weakened by the long-term effects of microgravity. To keep fit, it has been suggested that they can do some exercise using a Space Cycle: a horizontal beam from which an exercise bike and a cage are suspended. One astronaut sits on the exercise bike and pedals, which causes the whole Space Cycle to rotate around a pole. Another astronaut standing in the cage experiences artificial gravity. When rotated at 20 revolutions per minute, this is of similar strength to the gravitational field on Earth.

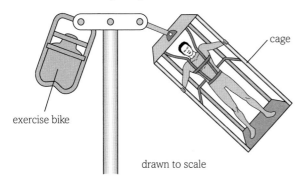

space cycle

cage

exercise bike

drawn to scale

(a) Calculate the angular velocity, in rad s^{-1}, corresponding to 20 revolutions per minute. [2]

(b) Use the diagram to estimate the radius of the path followed by the cage's platform and use this estimate to calculate the platform's acceleration. [3]

(Total for Question 6 = 5 marks)

7 The London Eye is a ferris wheel. It has a large vertical circle with 32 equally-spaced passenger cabins attached to it. The wheel rotates so that each cabin has a constant speed of 0.26 m s^{-1} and moves around a circle of radius 61 m.

(a) Calculate the time taken for each cabin to make one complete revolution. [2]

(b) Calculate the centripetal force acting on each cabin. Mass of cabin = 9.7×10^3 kg [2]

(c) (i) The diagram shows just the circle and the cabins. Draw arrows to show the direction of the centripetal force acting on a person in a cabin when the person is at each of positions **A**, **B** and **C**. [1]

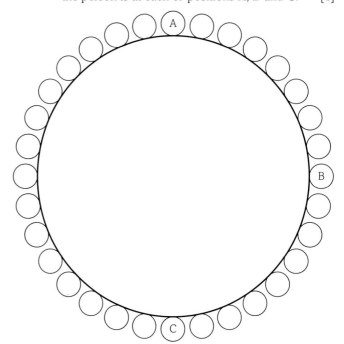

*(ii) As the person in a cabin moves around the circle, the normal contact force between the person and the cabin varies.
State the position at which this force will be a maximum and the position at which it will be a minimum. Explain your answers. [4]

(Total for Question 7 = 9 marks)

8 Describe an experiment you could carry out in a school laboratory to verify the equation for the centripetal force:

$$F = \frac{mv^2}{r}$$

Include details of the measurements to be made, and the precautions that would minimise experimental errors. Also include explanations of the analysis of the experimental results that would confirm the verification. [6]

(Total for Question 8 = 6 marks)

TOPIC 6 ELECTRIC AND MAGNETIC FIELDS

CHAPTER

6A   ELECTRIC FIELDS

The usual explanation for the spiky hair caused by touching a Van de Graaff generator is that the excess charges it puts on the body all repel each other, so each hair experiences a force pushing it away from the others and from the head. Whilst this is true, the explanation is too simple. The hair shows us the existence of an electric field coming from the person's head, with the hair trying to follow the electric field lines. Gravity also affects the hair, so there is a resultant force effect. If your school has a Van de Graaff generator, you could try setting up an experiment where the person in contact with it is in different positions (i.e. lying down on a wooden bench, bending over at the waist, etc.), so that the electrical and gravitational forces are not working against each other quite so much!

In this chapter, we will see the difference between uniform and radial fields, and learn how to calculate the strength of both types of electric field. We will also explore the concept of electric potential, which relates to the storage and delivery of electrical energy.

MATHS SKILLS FOR THIS CHAPTER

- Recognising and use of appropriate units in calculations (*e.g. comparing* $N\,C^{-1}$ *and* $V\,m^{-1}$)

- Use of trigonometric functions (*e.g. in verifying Coulomb's law from angled strings*)

- Identifying uncertainties in measurements (*e.g. the error in the conclusions from Millikan's oil drop experiment*)

- Substitute numerical values into algebraic equations using appropriate units for physical quantities (*e.g. in calculations of radial electric field strength*)

- Changing the subject of an equation (*e.g. finding the separation of two charges from Coulomb's law*)

- Plotting two variables from experimental data (*e.g. from a Coulomb's law experiment*)

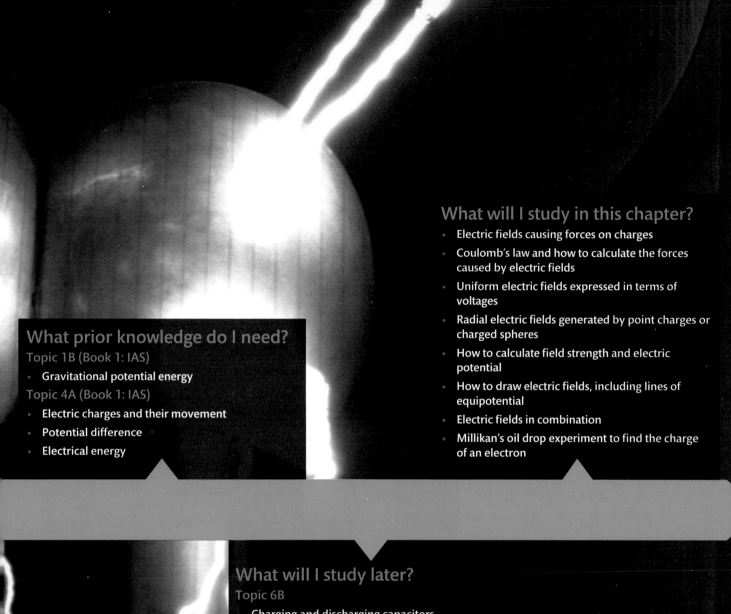

What will I study in this chapter?

- Electric fields causing forces on charges
- Coulomb's law and how to calculate the forces caused by electric fields
- Uniform electric fields expressed in terms of voltages
- Radial electric fields generated by point charges or charged spheres
- How to calculate field strength and electric potential
- How to draw electric fields, including lines of equipotential
- Electric fields in combination
- Millikan's oil drop experiment to find the charge of an electron

What prior knowledge do I need?

Topic 1B (Book 1: IAS)

- Gravitational potential energy

Topic 4A (Book 1: IAS)

- Electric charges and their movement
- Potential difference
- Electrical energy

What will I study later?

Topic 6B

- Charging and discharging capacitors

Topic 6C

- Generating electricity

Topic 7A

- Alpha particle scattering by nuclear electric fields

Topic 7B

- Accelerating subatomic particles using electric fields

Topic 11A

- Comparisons with gravitational fields

6A | 1 ELECTRIC FIELDS

LEARNING OBJECTIVES

- Define electric field and electric potential.
- Use the equation $E = \dfrac{F}{Q}$ for electric field strength.
- Describe the concept of a uniform electric field.
- Use the equation $E = \dfrac{V}{d}$ for uniform electric field strength between parallel plates.

PUSHING CHARGES

Many machines work by using fast-moving charged particles. For example, in a hospital X-ray machine, high speed electrons are crashed into a metal target in order to produce the X-rays. So how do we cause the electrons to move at high speed? A space that will cause charged particles to accelerate is said to have an **electric field**.

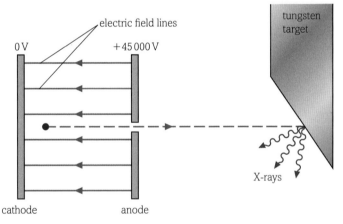

fig A An electric field accelerates electrons in an X-ray machine.

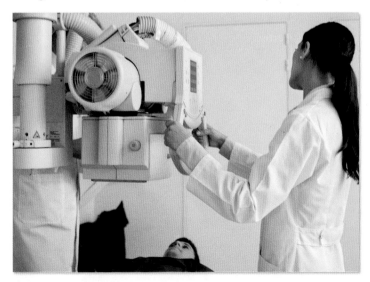

fig B An X-ray machine.

A force will act on a charged particle when it is in an electric field. To visualise the forces caused by the field, we draw **electric field lines**. These show the direction in which a positively charged particle will be pushed by the force that the field produces. Lines which are closer together show a stronger field. This is the same for all field patterns, including magnetic fields. A stronger field causes stronger forces.

The force, F, that acts on a charged particle is the **electric field strength** (E) multiplied by the amount of charge in coulombs (Q), as given by the equation:

$$F = EQ$$

From this force equation, we can also see how quickly a charge would accelerate. Newton's second law is that $F = ma$, so we can equate the two equations:

$$F = EQ = ma$$

So: $\quad a = \dfrac{EQ}{m}$

WORKED EXAMPLE 1

(a) What force will an electron experience when it is in an electric field generated by an X-ray machine's electric field which has a strength of $4.5 \times 10^5\,\mathrm{V\,m^{-1}}$?

$$F = EQ$$
$$= 4.5 \times 10^5 \times -1.6 \times 10^{-19}$$
$$F = -7.2 \times 10^{-14}\,\mathrm{N}$$

The minus sign indicates that the electron will feel a force trying to accelerate it towards the more positive end of the field. This is the opposite direction to the conventional field direction which, like electric current, goes from positive to negative.

(b) How fast will the electron be travelling if this field accelerates it from rest and it is within the field for a distance of 10 cm?

The kinematics equations of motion have $v^2 = u^2 + 2as$. 'From rest' means that $u = 0$, so:

$$v = \sqrt{2as} = \sqrt{\dfrac{2EQs}{m}}$$
$$= \sqrt{\dfrac{2 \times (4.5 \times 10^5) \times (-1.6 \times 10^{-19}) \times (-0.1)}{(9.11 \times 10^{-31})}}$$
$$v = -1.26 \times 10^8\,\mathrm{m\,s^{-1}}$$

Again, the minus sign indicates that the motion is in the opposite direction to the electric field, i.e. towards the more positive end of it. Relativistic effects are ignored in this example.

ELECTRIC POTENTIAL

In the worked example above, the kinetic energy gained by the electron came from a transfer of electrical potential energy. The electron had this energy due to its location within the electric field. Every location within a field gives a charged particle a certain electric potential energy per unit of charge. This is called **electric potential**, $V = \dfrac{E_p}{Q}$. This is similar to the relationship between force and field strength: field strength is the force that a charged particle experiences per unit of charge, expressed by the equation $E = \dfrac{F}{Q}$.

The difference between that and a new location that the electron might move to is called the **potential difference** which the electron moves through. You may remember (**Book 1, Section 4A.2**) that we previously defined potential difference for a device in an electrical circuit as the energy transferred per coulomb of charge passing through the device. In an electric field, we can follow exactly the same idea to find out how much kinetic energy a charged particle will gain by moving within the field. This is given by the equation:

$$E_k = VQ$$

WORKED EXAMPLE 2

(a) What is the kinetic energy of the electron in part (b) of the previous worked example above?

$$E_k = \tfrac{1}{2}mv^2$$
$$= \tfrac{1}{2} \times (9.11 \times 10^{-31}) \times (1.26 \times 10^8)^2$$
$$E_k = 7.2 \times 10^{-15}\ \text{J}$$

(b) What is the kinetic energy gained by an electron as it is accelerated through a potential difference of 45 kV?

$$E_k = VQ = 45 \times 10^3 \times 1.6 \times 10^{-19}$$
$$E_k = 7.2 \times 10^{-15}\ \text{J}$$

These two answers are the same because they are actually calculations of the same thing.

UNIFORM FIELDS

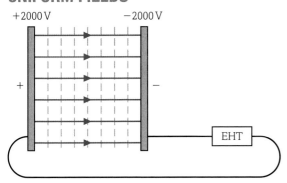

+2000 V   −2000 V

EHT

▲ **fig C** The blue arrows show the direction of the uniform electric field produced between parallel plates that have a potential difference between them. The dashed lines show where the potential is constant along equipotential lines.

An electric field exists between any objects which are at different electrical potentials. So, if we connect parallel metal plates to a power supply, we can set up a potential difference, and therefore an electric field, between them. This is shown in **fig C**. The field is uniform if its field lines are parallel across the whole field.

The strength of a uniform electric field is a measure of how rapidly it changes the potential over distance. The equation which describes this divides the potential difference, V, by the distance over which the potential difference exists, d:

$$E = \dfrac{V}{d}$$

WORKED EXAMPLE 3

In the X-ray machine shown in **fig A**, there is a potential difference of 45 000 V between a cathode and an anode. These electrode plates are 10 cm apart. What is the electrical field strength between the plates?

$$E = \dfrac{V}{d} = \dfrac{45\,000}{0.1}$$
$$E = 4.5 \times 10^5\ \text{V m}^{-1}$$

EXAM HINT

Be careful not to confuse E for energy, with E for electric field.

DID YOU KNOW?

Some medical X-ray machines can generate voltages up to 400 000 V.

PRACTICAL SKILLS

Investigating electric fields

▲ **fig D** Finding electric field lines in the lab.

Make sure you know how to use a high voltage power supply safely before conducting this experiment. Only use a high voltage power supply designed for school experiments.

You can investigate the shapes of electric fields using a high voltage power supply to provide a potential difference, and any clear oil with floating small, dry seeds to show the field lines. The small, dry seeds become slightly charged, and the charges on the seeds cause them to line up. This shows the lines of action of the forces produced by the field. Try it with different shaped electrodes to see uniform and non-uniform fields.

Safety Note: High voltage power supplies can give a severe electric shock and should be used with shrouded connectors. Do not touch any switches or sockets with wet hands.

EQUIPOTENTIALS

As we move through an electric field, the electrical potential changes from place to place. Those locations that all have the same potential can be connected by lines called **equipotentials**. These are very much like the contours on a map, which are lines of equal height, but here they show lines of gravitational equipotential. The field will always be perpendicular to the equipotential lines, as a field is defined as a region which changes the potential. How close the equipotentials are indicates the strength of the electric field, as this shows how rapidly the electric potential is changing.

PRACTICAL SKILLS

Investigating equipotentials

▲ **fig E** Apparatus for investigating equipotentials.

You can investigate how electrical potential varies across an electric field by measuring the voltage between zero and the point within the field that you are interested in. A simple experiment allows you to show where equipotentials are and enables you to produce a picture of the electric field. In this you arrange conducting paper and a low voltage power supply to create an electric field. Try it with differently shaped electrodes to see how they change the field shape.

MILLIKAN'S OIL DROP EXPERIMENT

Scientists are very interested to find out what the fundamental particles are and how they behave. They can use a mass spectrometer to investigate this and find the ratio of charge/mass for particles. This is a fundamental property of a charged particle and each one is different. However, it was not possible to find the mass of an electron until the value of its charge had been measured. In 1909, Robert Millikan developed an experiment (see **fig F**) which determined the charge on a single electron.

Although Millikan had a variety of extra equipment to make his experiment work successfully, the basic idea of the experiment is very simple. The weight of a charged droplet of oil is balanced by the force from a uniform electric field. (Millikan also did a variation in which there is no field and the terminal velocity is measured, and then one in which the field provides a stronger force than gravity and the terminal velocity upwards is measured.) When oil is squirted into the upper chamber from the vaporiser, friction gives the droplets an electrostatic charge. This will be some (unknown) multiple of the charge on an electron, because electrons have been added or removed due to friction. As the drops fall under gravity, some will go through the anode and enter the uniform field created between the charged plates. If the field is switched off, they will continue to fall at their terminal velocity (see **fig G**).

▲ **fig F** Millikan's oil drop apparatus for finding the charge on an electron. The chamber shown was suspended in a trough of motor oil to reduce heat transfer within it, and so reduce any problems of convection currents affecting falling oil droplets within the chamber.

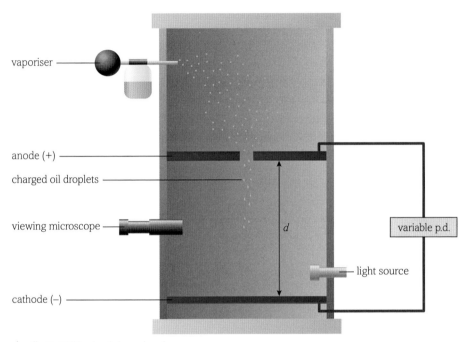

▲ **fig G** Millikan's oil drop chamber.

> **LEARNING TIP**
>
> **Stokes' law**
> Sir George Gabriel Stokes investigated fluid dynamics and derived an equation for the viscous drag (F) on a small sphere moving through a fluid at low speeds:
>
> $$F = 6\pi\eta vr$$
>
> where r is the radius of sphere, v is the velocity of sphere, and η is the coefficient of viscosity of the fluid.

For Millikan's oil drops, the density of air in the chamber is so low that its upthrust is generally insignificant (although we would have to consider it if we wanted to do really accurate calculations). Under terminal velocity, the weight equals the viscous drag force:

$$mg = 6\pi\eta v_{term} r$$

where η is the viscosity of air, and r is the radius of the drop.

When held stationary by switching on the electric field and adjusting the potential, V, until the drop stands still:

weight = electric force

$$mg = EQ = \frac{VQ}{d}$$

By eliminating the weight from the two situations, this means that:

$$6\pi\eta v_{term} r = \frac{VQ}{d}$$

or: $$Q = \frac{6\pi\eta v_{term} rd}{V}$$

Millikan could not measure r directly, so he had to eliminate it from the equations. Further development of Stokes' law tells us that a small drop falling at a low terminal velocity will follow the equation:

$$v_{term} = \frac{2r^2 g(\rho_{oil} - \rho_{air})}{9\eta}$$

which, if we again ignore the low density of air, rearranges to:

$$r = \left(\frac{9\eta v_{term}}{2g\rho_{oil}}\right)^{\frac{1}{2}}$$

Overall, then:

$$Q = \frac{6\pi\eta v_{term}d}{V} \times \left(\frac{9\eta v_{term}}{2g\rho_{oil}}\right)^{\frac{1}{2}}$$

Millikan did the experiment several hundred times, including repeated measurements on each drop. Many times he let the drop fall, before halting it with an electric field, and then lifting it up again with a stronger field, before letting it fall again. From this data, he found that the charges on the droplets were always a multiple of 1.59×10^{-19} C, which is less than 1% away from the currently accepted value of 1.602×10^{-19} C.

CHECKPOINT

SKILLS ADAPTIVE LEARNING

1. ▶ What is the force on an electron in an electric field of 300 V m⁻¹?

2. What is the strength of an electric field that will put a force of 1.28×10^{-15} N on a proton?

3. How much will an alpha particle accelerate while it is in an electric field of 10 kV m⁻¹?

4. ▶ In the electron beam of a cathode ray oscilloscope, electrons are accelerated through a potential difference of 3000 V. This potential difference is set up between parallel plate electrodes that are 3 cm apart.
   (a) Calculate the electric field strength between these electrodes. Assume it is a uniform field.
   (b) How much faster will the electrons be moving when they emerge from this field?
   (c) Draw a diagram to illustrate the field produced by these plates.
   (d) Explain how the effects of the field would be different if a proton were placed in it.

SKILLS INTERPRETATION

5. **Fig H** shows a positively charged oil drop held at rest between two parallel conducting plates, A and B.

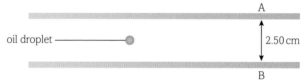

oil droplet ───────●   A
                                              ↕ 2.50 cm
                                              B

▲ **fig H**

   (a) The oil drop has a mass of 9.79×10^{-15} kg. The potential difference between the plates is 5000 V and plate B is at a potential of 0 V. Is plate A positive or negative?
   ▶ (b) Draw a labelled free-body force diagram which shows the forces acting on the drop.
   (c) Calculate the electric field strength between the plates.
   (d) Calculate the magnitude of the charge Q on the drop.
   (e) How many electrons would have to be removed from a neutral oil drop for it to have this charge?

SUBJECT VOCABULARY

electric field a region of space that will cause charged particles to experience a force
electric field lines imagined areas where the electric field has an effect
electric field strength the force-to-charge ratio for a charged particle experiencing a force due to an electric field
electric potential a measure of possible field energy in a given location within that field; the energy per unit charge at that location
potential difference the change in potential between two locations in a given field
equipotentials positions within a field with zero potential difference between them

LEARNING OBJECTIVES

- Define radial electric fields.
- Draw and interpret diagrams of electric fields.
- Describe the relationship between electric field strength and electric potential.
- Use the equations relating to field strength and potential for radial electric fields.

RADIAL FIELDS

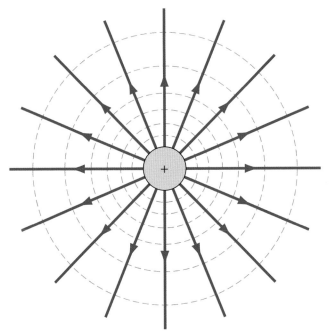

▲ **fig A** The radial electric field around a proton (or any positively charged sphere).

In the region around a positively charged sphere, or point charge such as a proton, the electric field will act outwards in all directions away from the centre of the sphere, as shown in **fig A**. You will see that the arrows in **fig A** get further apart as you move further away from the sphere. This indicates that the field strength reduces as you move away from the centre. The distance between equipotentials also increases as you move further away. The field is the means by which the potential changes, so if it is weaker, the potential changes less quickly.

A charged particle inside a charged sphere would experience no resultant force and so there is no electric field. The overall effect of all the charges on the sphere cancel out within the sphere itself. Distance measurements for use in calculations should always be considered from the centre of a charged sphere.

RADIAL ELECTRIC FIELD STRENGTH

We have seen that the field lines for a radial field become further apart as you go away from the centre. This means that the equation for the field strength of a radial electric field must incorporate this weakening of the field with distance from the charge. The expression for radial field strength at a distance r from a charge Q is:

$$E = \frac{Q}{4\pi\varepsilon_0 r^2}$$

This is only truly correct for a field that is produced in a vacuum. This is because the value ε_0 – the permittivity of free space – is a constant which relates to the ability of the fabric of the Universe to support electric fields. Its value is $\varepsilon_0 = 8.85 \times 10^{-12}\,\mathrm{F\,m^{-1}}$. Other substances, for example, water, may be better or worse at supporting electric fields, so an extra factor (the relative permittivity, ε_r) would appear in the equation to account for this. Air is considered to be near enough a vacuum that we use the equation given above for electric fields in air.

WORKED EXAMPLE 1

What is the electric field strength at a distance of 1 angstrom ($1 \times 10^{-10}\,\mathrm{m}$) from a proton?

$$E = \frac{Q}{4\pi\varepsilon_0 r^2}$$

$$= \frac{(1.6 \times 10^{-19})}{4 \times 3.14 \times 8.85 \times 10^{-12} \times (1 \times 10^{-10})^2}$$

$$E = 1.44 \times 10^{11}\,\mathrm{V\,m^{-1}}$$

POTENTIAL IN A RADIAL ELECTRIC FIELD

Electric field strength tells us how quickly the electric potential is changing. A stronger field will have the equipotentials closer together. This equation states that the electric field strength, E, is equal to the rate of change of potential, V, with distance, x:

$$E = -\frac{dV}{dx}$$

This leads to the expression for radial field potential at a distance r from a charge Q:

$$V = \frac{Q}{4\pi\varepsilon_0 r}$$

You will see from the equation above that the position of zero potential can be considered to be at infinity.

WORKED EXAMPLE 2

What is the electric potential at a distance of 1 angstrom ($1 \times 10^{-10}\,\mathrm{m}$) from a proton?

$$V = \frac{Q}{4\pi\varepsilon_0 r}$$

$$= \frac{(1.6 \times 10^{-19})}{4 \times 3.14 \times 8.85 \times 10^{-12} \times (1 \times 10^{-10})}$$

$$V = 14.4\,\mathrm{V}$$

The constant expression $\frac{1}{4\pi\varepsilon_0}$ is sometimes written as the single letter, k (Coulomb's constant). The value for k is:

$$k = \frac{1}{4\pi\varepsilon_0} = 8.99 \times 10^9 \text{ N m}^2 \text{ C}^{-2}$$

This means our expressions for radial field strength and potential can be written:

$$E = \frac{kQ}{r^2}$$

$$V = \frac{kQ}{r}$$

COMBINING ELECTRIC FIELDS

When there are electric fields caused by more than one charged object, the overall field at any place is the vector sum of all the contributions from each field. The force effects of each contributing field combine to produce the overall field. At every point you have to work out the resultant force and so you know how a charged particle will be affected in that place. The sum of all of these individual force effects is the overall electric field.

Charge is particularly concentrated in regions around spikes or points on charged objects. This means that they have close field lines and the field will be strong around them. For this reason, lightning conductors are spiked: the concentrated charge will more strongly attract opposite charges, so lightning is more likely to hit the lightning conductor than the building being protected.

In fact, the field around a spiked lightning conductor is often so strong that it can cause charge leakage through the conductor. This happens before charge builds up enough to produce a lightning strike. Therefore, the probability of lightning occurring is reduced, further protecting the building from these dangers.

(a)   **(b)**

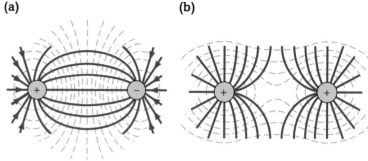

(c)   **(d)**

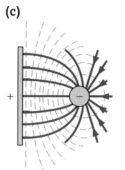 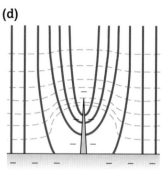

▲ **fig B** Complex fields also always have their equipotentials perpendicular to the field lines at all points.

CHECKPOINT

SKILLS   PROBLEM-SOLVING

1. ▶ A Van de Graaff generator is charged with 2.5×10^{-7} coulombs in a school lab demonstration. The spherical metal dome has a diameter of 30 cm.

   (a) Draw the electric field around the charged dome of the generator.

   (b) Add at least three equipotential lines to your diagram. Label these with their potential values, following the correct scale of your diagram.

   (c) What is the electric field strength at a distance of 5 cm from the surface of the charged dome?

   (d) What is the electric potential at a distance of 35 cm from the surface of the charged dome?

2. Explain what would happen to an electron at the exact centre of the electric field shown in **fig B(b)**.

3. Draw a pair of electrons 8 cm apart. On your diagram, add field lines to show the shape of the electric field produced by these two electrons. Also add in several equipotential lines.

4. Why is the electric field in **fig B(d)** strongest near the point of the spike?

5. If **fig B(a)** shows a hydrogen atom, then the separation of the charges would be 5.3×10^{-11} m.

   (a) What is the strength of the electric field caused by the proton, at the point where the electron is?

   (b) Calculate the force on the electron caused by the proton's electric field at this distance.

EXAM HINT

When a question uses the command word **Explain** (see question 2 above), your answer must contain some reasoning or justification, which can include mathematical explanations.

CHARGED PARTICLE INTERACTIONS

The attraction between a proton and an electron can be imagined as the proton creating an electric field because of its positive charge, and the electron experiencing a force produced by the proton's field. (Note that it could also be thought of the other way round, with the proton experiencing a force caused by the electron's electric field.)

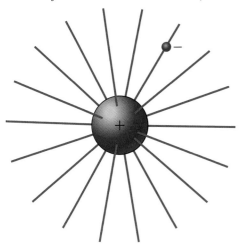

▲ **fig A** A proton's field will cause a force on a negatively charged electron.

The force between two charged particles, Q_1 and Q_2, which are separated by a distance r, is described by Coulomb's law and is given by the expression:

$$F = \frac{Q_1Q_2}{4\pi\varepsilon_0 r^2} = \frac{kQ_1Q_2}{r^2}$$

This can be seen as a logical consequence of the equations we have already seen for electric fields:

$$E = \frac{Q}{4\pi\varepsilon_0 r^2}$$

So, for one charge Q_1 creating a radial field in which another charge Q_2 sits at a distance r from Q_1:

$$F = EQ = \frac{Q_1Q_2}{4\pi\varepsilon_0 r^2}$$

Note that for charged spheres, the calculation of the force between them is also given by Coulomb's law, but the distance is measured between the centres of the spheres.

WORKED EXAMPLE 1

What is the force of attraction between the electron and proton in a hydrogen atom?

$r = 5.3 \times 10^{-11}$ m $Q_1 = 1.6 \times 10^{-19}$ C $Q_2 = -1.6 \times 10^{-19}$ C

$$F = \frac{Q_1Q_2}{4\pi\varepsilon_0 r^2}$$

$$= \frac{(8.99 \times 10^{19}) \times (1.6 \times 10^{-19}) \times (-1.6 \times 10^{-19})}{(5.3 \times 10^{-11})^2}$$

$F = -8.2 \times 10^{-8}$ N

PRACTICAL SKILLS

Investigating Coulomb's law

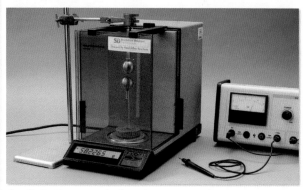

▲ **fig B** Measuring the force between two charges.

A pair of insulated metal spheres can be charged using a Van de Graaff generator, or simply by induction using a charged plastic rod. Clamping the spheres close to each other will cause a force between them which can be measured using an electronic balance. By changing the distance of separation and measuring the force at each distance, this apparatus can be used to confirm that Coulomb's law follows an inverse square law. See also an alternative method for verifying Coulomb's law below.

Safety Note: A Van de Graaff generator will give an electric shock but not big enough to harm a normal healthy person. A person with a heart problem could be badly affected and even have a heart attack.

EVERYDAY LEVITATION

All materials are made of atoms, and the outer part of atoms is believed to be electron orbits (negative charge). We can therefore imagine the top of any surface to consist of a sheet of negative charges. So, the surface of the floor and the bottom of a man's shoe will both be covered with negative charge. The repulsion between these two surfaces will cause the man to levitate ever so slightly above the floor, not actually touching it.

WORKED EXAMPLE 2

Estimate the height at which a man levitates.

Estimate of man's weight = 800 N

Therefore each shoe supports 400 N.

Estimate of shoe surface area = 10 cm by 30 cm

Area of shoe surface = $0.1 \times 0.3 = 0.03 \, \text{m}^2$

Estimate of average diameter of atoms in each surface, $d_{atom} = 1.2 \times 10^{-10} \, \text{m}$

Cross-sectional area of each atom = πr^2_{atom}

$A_{atom} = \pi(0.6 \times 10^{-10})^2$

$A_{atom} = 1.13 \times 10^{-20} \, \text{m}^2$

Number of atoms in bottom surface of shoe = $\dfrac{\text{area of shoe}}{\text{area one atom}}$

$N = \dfrac{0.03}{1.13 \times 10^{-20}} = 2.65 \times 10^{18}$ atoms

Force supported by each shoe atom = $\dfrac{\text{force supported by shoe}}{N}$

$F_{atom} = \dfrac{400}{2.65 \times 10^{18}} = 1.51 \times 10^{-16} \, \text{N}$

Assume each shoe atom electron repels a corresponding one in the floor surface, and oxygen, with eight electrons, is the most abundant element. So estimate the force per atom that is shared between four electrons (with the other four being on the opposite side of the atom at any given moment).

$$F_{electron} = \frac{F_{atom}}{4} = 3.77 \times 10^{-17} \, \text{N}$$

$F_{electron}$ = coulomb force repelling shoe electron and floor electron

$$= \frac{Q_1 Q_2}{4\pi\varepsilon_0 r^2}$$

$$r = \sqrt{\frac{Q_1 Q_2}{4\pi\varepsilon_0 F_{electron}}} = \sqrt{\frac{(-1.6 \times 10^{-19}) \times (-1.6 \times 10^{-19})}{4 \times 3.14 \times (8.85 \times 10^{-12}) \times (3.77 \times 10^{-17})}}$$

Levitation height: $r = 2.5 \times 10^{-6} \, \text{m}$

So the man will float a couple of micrometres above the floor surface; a distance equal to about a tenth of the size of a human red blood cell, or the size of a bacterium such as *Escherichia coli*. There are a number of assumptions in this calculation, especially that the positive charge of the nuclei of the atoms involved has not been considered. Also, the compression of atomic bonds within each material will mean that the force is supported across more atoms than just those considered. However, it is true that the repulsion of atomic electrons provides most of the contact forces that we experience in our lives. This force increases with the square of a decreasing separation and does mean that you can never actually touch anything.

PRACTICAL SKILLS

Verifying Coulomb's law

In 1785, Charles Augustine de Coulomb published a paper in which he provided experimental evidence for the equation to calculate the force between two charges. His experiments involved measuring the force between charged spheres on a torsion (rotating) balance and fixed charged spheres. A simplification of his method, which can be carried out in a school lab, is shown in **fig C**.

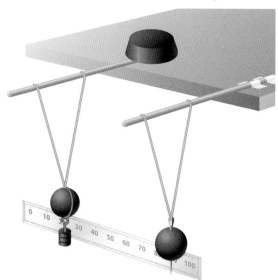

▲ **fig C** A simple method to verify Coulomb's law.

By reducing the separation of the spheres, you will see an increase in the angle of the lighter hanging ball. You can investigate the force between charged spheres and confirm that there is an inverse square law involved, and that the force is proportional to the product of the charges.

Fig D shows the forces on the hanging ball. By resolving and equating the horizontal and vertical components, we can develop a method for verifying Coulomb's law.

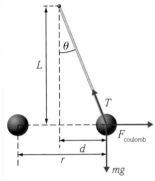

▲ **fig D** Forces on the hanging charged ball.

The horizontal component of the tension balances with the Coulomb force:

$$F_{coulomb} = T\sin\theta$$

And the weight balances the vertical component of tension:

$$mg = T\cos\theta$$

Dividing the two equations gives:

$$\frac{F_{coulomb}}{mg} = \frac{T\sin\theta}{T\cos\theta} = \tan\theta$$

$$\tan\theta = \frac{d}{L}$$

$$\frac{F_{coulomb}}{mg} = \frac{d}{L}$$

$$F_{coulomb} = \frac{mgd}{L}$$

$$\frac{Q_1Q_2}{4\pi\varepsilon_0 r^2} = \frac{mgd}{L}$$

For a fixed amount of charge on the spheres, you could vary r and measure d, from which a plot of $\frac{1}{r^2}$ against d should produce a straight line. It is very difficult to control the amount of charge placed on the spheres. However, if you charge them both and then touch one sphere with a third sphere which is uncharged, you will remove half of the charge on that one. Do this several times, and measure d in each case. This will enable you to confirm that $F_{coulomb}$ is proportional to the product of the two charges.

CHECKPOINT

SKILLS ANALYSIS

1. ▶ What is the force of attraction between a uranium nucleus (atomic number 92) and an electron at a distance of 0.1 nm?

2. (a) What is the strength of the electric field caused by a gold nucleus (atomic number 79) at a distance of 1×10^{-12} m (1 pm) from the centre of the nucleus?

    (b) What is the force of repulsion between an alpha particle and a gold nucleus when the alpha passes by the nucleus at a distance of 1 pm?

3. Explain a potential problem with the experimental set-up in **fig C** when comparing with Coulomb's law that $F = \dfrac{Q_1Q_2}{4\pi\varepsilon_0 r^2}$

    (Hint: what would your ruler be measuring?)

4. Estimate the distance that you float above your bed at night. Consider the Coulomb repulsion between the electrons in the bed and those in you.

5. ▶ Jamal and Izaac undertook an experiment to test Coulomb's law, following the method of **fig C**. The measurements they took are given in **table A**.

d / m	r / m
0.002	0.036
0.004	0.028
0.006	0.024
0.008	0.021
0.010	0.019
0.012	0.018
0.014	0.016
0.016	0.014

table A Experimental results from a Coulomb's law experiment.

Draw an appropriate graph to analyse their results to show that the Coulomb force follows an inverse square relationship.

EXAM HINT

When a question uses the command word **Estimate** (see question 4 above) there is no exact answer. You will usually need to make a calculation using reasonable guesses at the values needed to make that calculation. It is a good idea to start your answer by stating all the values you have chosen for the quantities needed for the calculation.

POWER LINE PROBLEMS

SKILLS ▸ CRITICAL THINKING, PROBLEM SOLVING, INTERPRETATION, CONTINUOUS LEARNING, PRODUCTIVITY, COMMUNICATION

In 2010, a 500 kV high voltage transmission line was proposed to carry electricity for 28 miles across Washington State in the USA. The Bonneville Power Administration commissioned various reports to investigate safety and other public interest issues related to the new power lines, to inform local government officers. Dan Bracken, an independent scientific consultant, wrote a substantial report on the impact of electric fields that might be generated by the power lines.

SCIENTIFIC REPORT

ELECTRIC FIELDS AND LIVING TISSUE

An electric field is said to exist in a region of space if an electrical charge, at rest in that space, experiences a force of electrical origin (i.e. electric fields cause free charges to move). Electric field is a vector quantity: that is, it has both magnitude and direction. The direction corresponds to the direction that a positive charge would move in the field. Sources of electric fields are unbalanced electrical charges (positive or negative) and time-varying magnetic fields. Transmission lines, distribution lines, house wiring, and appliances generate electric fields in their vicinity because of the unbalanced electrical charges associated with voltage on the conductors. On the power system in North America, the voltage and charge on the energized conductors are cyclic (plus to minus to plus) at a rate of 60 times per second. This changing voltage results in electric fields near sources that are also time-varying at a frequency of 60 hertz (Hz; a frequency unit equivalent to cycles per second).

As noted earlier, electric fields are expressed in units of volts per metre (V/m) or kilovolts (thousands of volts) per metre (kV/m).

The spatial uniformity of an electric field depends on the source of the field and the distance from that source. On the ground, under a transmission line, the electric field is nearly constant in magnitude and direction over distances of several feet (1 metre). However, close to transmission- or distribution-line conductors, the field decreases rapidly with distance from the conductors. Similarly, near small sources such as appliances, the field is not uniform and falls off even more rapidly with distance from the device. If an energized conductor (source) is inside a grounded conducting enclosure, then the electric field outside the enclosure is zero, and the source is said to be shielded.

Electric fields interact with the charges in all matter, including living systems. When a conducting object, such as a vehicle or person, is located in a time-varying electric field near a transmission

▲ **fig A** Transmission lines can carry very high voltages, generating significant electric fields.

line, the external electric field exerts forces on the charges in the object, and electric fields and currents are induced in the object. If the object is grounded, then the total current induced in the body (the 'short-circuit current') flows to earth. The distribution of the currents within, say, the human body, depends on the electrical conductivities of various parts of the body: for example, muscle and blood have higher conductivity than bone and would therefore experience higher currents.

At the boundary surface between air and the conducting object, the field in the air is perpendicular to the conductor surface and is much, much larger than the field in the conductor itself. For example, the average surface field on a human standing in a 10 kV/m field is 27 kV/m; the internal fields in the body are much smaller: approximately 0.008 V/m in the torso and 0.45 V/m in the ankles.

Section 3.1 from Appendix E, Electrical Effects, of '*BIG EDDY – KNIGHT 500-kV TRANSMISSION PROJECT*', by T. Dan Bracken, for Bonneville Power Administration, dated March 2010, *https://www.bpa.gov/efw/Analysis/NEPADocuments/nepa/Big_Eddy-Knight/AppendixE-ElectricalEffectsCombined.pdf*

SCIENCE COMMUNICATION

1 (a) Explain why the author has explained electric fields from first principles in this report.

   (b) Why has he still used a significant amount of technical terminology and high level vocabulary?

2 The photograph shown here was not in the original report, but has been added for this textbook. Explain why there might be no pictures in the report produced by Mr Bracken.

3 What is the importance in this report of including references to the effects of electric fields on living things?

INTERPRETATION NOTE

Think about who is the intended audience for this information.

PHYSICS IN DETAIL

Now we will look at the physics in detail. Some of these questions will link to topics elsewhere in this book, so you may need to combine concepts from different areas of physics to work out the answers.

4 (a) As described, the voltage changes from positive to negative and back again, like a wave cycle. What is the time period for a cycle of the voltage in the USA?

   (b) Why would the author have chosen to express field values in units of $V\,m^{-1}$ rather than the other correct possibility of $N\,C^{-1}$?

5 (a) We have learned that the electric field around point charges is radial. Why could the field coming out of a long straight cable also be referred to as radial?

   (b) Draw a diagram to show a high voltage power cable, 10 m above a human standing on the ground. Add electric field and equipotential lines to your diagram to show how it emerges from the cable, and connects to the ground. Assume a fixed instant in time, when the voltage in the cable is exactly 500 kV.

6 (a) What does this extract tell us about the comparative resistivities of blood and bone?

   (b) Explain why you think this difference exists.

PHYSICS TIP

The proposed 500 kV cable would be at a minimum height of 11 m.

PHYSICS TIP

Consider the differences in the substances making up blood and bone.

ACTIVITY

The author has written that 'close to transmission- or distribution-line conductors, the field decreases rapidly with distance from the conductors.' Imagine that the senators from Washington State government have written a letter to ask for an explanation of this claim. Write a letter in response which explains the mathematics involved in radial electric fields. Include some example calculations to illustrate the values involved, for comparison with other values already given in the report.

1 A correct unit for electric field strength is:

A $C\,m^{-1}$

B $kV\,m^{-1}$

C $N\,C$

D $V\,m$ [1]

(Total for Question 1 = 1 mark)

2 Which of these particles is **not** affected by an electric field?

A electron

B neutron

C proton

D Cu^{2+} ion [1]

(Total for Question 2 = 1 mark)

3 The electric field strength from a negative point charge:

A decreases in magnitude with the square of the charge

B decreases in magnitude with the square of the distance from the charge

C increases in magnitude with distance from the charge

D increases in magnitude with the square of the distance from the charge. [1]

(Total for Question 3 = 1 mark)

4 The force between two protons held 1 nm apart is:

A $2.6 \times 10^{-20}\,N$

B $2.3 \times 10^{-19}\,N$

C $2.3 \times 10^{-10}\,N$

D $1.44 \times 10^{9}\,N$ [1]

(Total for Question 4 = 1 mark)

5 Which row in the table correctly shows the properties of both uniform and radial electric fields?

	Uniform fields	Radial fields
A	Field lines + to −	Equipotentials equally spaced
B	Field lines + to −	Field lines equally spaced
C	Field lines equally spaced	Equipotentials equally spaced
D	Equipotentials equally spaced	Field lines + to −

[1]

(Total for Question 5 = 1 mark)

6 (a) Explain what is meant by a uniform electric field. [2]

(b) Describe how a uniform electric field can be demonstrated in a laboratory. [3]

(Total for Question 6 = 5 marks)

7 Two identical table tennis balls, M and N, are attached to non-conducting threads and suspended from a point P. The balls are each given the same positive charge and they hang as shown in the diagram. The mass of each ball is 2.7 g.

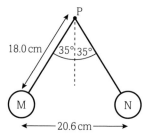

(a) Draw a free-body force diagram for ball M. Label your diagram with the names of the forces. [2]

(b) (i) Show that the tension in one of the threads is about 3×10^{-2} N. [3]

(ii) Show that the electrostatic force between the balls is about 2×10^{-2} N. [2]

(iii) Calculate the charge on each ball. [3]

(c) Explain what would have happened if the charge given to ball M was greater than the charge given to ball N. [2]

(Total for Question 7 = 12 marks)

8 Liquid crystal displays (LCDs) are made from two parallel glass plates, 10 μm apart, with liquid crystal molecules between them. The glass is coated with a conducting material.

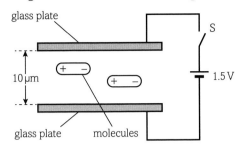

The molecules are positive at one end and negative at the other. They are normally aligned parallel with the glass plates, as shown.

The switch S is closed and 1.5 V is applied across the glass plates.

(a) Calculate the electric field strength between the plates. [2]

(b) Explain what happens to the liquid crystal molecules. [3]

(Total for Question 8 = 5 marks)

9 A charged sphere with a radius of 10 cm is positively charged to an electric potential of +1200 V.

(a) Calculate the charge on the sphere. [2]

(b) (i) Calculate the electric potential 35 cm from the surface of this sphere. [1]
    (ii) Calculate the electric potential 45 cm from the surface of this sphere. [1]

(c) An electron accelerates from rest at a position 45 cm from the surface of the sphere. Calculate the speed of the electron when it reaches a point 35 cm from the surface of the sphere. [3]

(Total for Question 9 = 7 marks)

10 The charge on an electron was originally measured in an experiment called the Millikan Oil Drop experiment.

In a simplified version of this experiment, an oil drop with a small electric charge is placed between two horizontal, parallel plates with a large potential difference (p.d.) across them. The p.d. is adjusted until the oil drop is stationary.

For a particular experiment, a p.d. of 5100 V was required to hold a drop of mass 1.20×10^{-14} kg stationary.

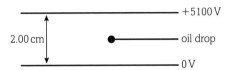

(a) Add to a copy of the diagram to show the electric field lines between the plates. [3]

(b) State whether the charge on the oil drop is positive or negative. [1]

(c) Complete the free-body force diagram to show the forces acting on the oil drop. You should ignore upthrust. [2]

(d) (i) Calculate the magnitude of the charge on the oil drop. [4]
    (ii) Calculate the number of electrons that would have to be removed or added to a neutral oil drop for it to acquire this charge. [2]

(Total for Question 10 = 12 marks)

11 With the aid of a diagram, explain how the electrical potential varies within the space around a stationary electron. [5]

(Total for Question 11 = 5 marks)

TOPIC 6 ELECTRIC AND MAGNETIC FIELDS

6B CAPACITORS

The touch screen on many smartphones works by measuring the capacitance of every point on a grid that covers the screen. Your finger conducts electricity, so when you touch the screen it changes the capacitance at that point. This is measured as a position on the grid of the screen so that the phone knows where you have touched and can follow your touch instruction. Capacitor-based touch screens cannot detect the touch if you wear non-conductive gloves, as the measurements are based on how well the points on the screen can store charge, and how charge can escape when you touch it.

The touch screen is one of the more interesting applications of the capacitor. However, as they can store energy, act as timing components, and smooth out varying currents, capacitors are found in almost all modern electronic devices.

In this chapter, we will see how electric fields drive electric charge to be stored on a capacitor, and how this simple process can lead to some complex but beautiful mathematics for the charge storage.

MATHS SKILLS FOR THIS CHAPTER

• **Recognising and use of appropriate units with prefixes** (*e.g. the microfarad*)

• **Use of calculators to work with exponential and logarithmic functions** (*e.g. charge remaining after a certain capacitor discharge time*)

• **Understanding the possible physical significance of the area between a curve and the x-axis and be able to calculate this area** (*e.g. finding the energy stored on a capacitor from a graph of voltage against charge*)

• **Changing the subject of an equation** (*e.g. finding the initial discharge current from the exponential equation*)

• **Interpreting logarithmic plots** (*e.g. plotting logarithmic data from a capacitor discharge experiment to find an unknown capacitance*)

• **Understanding that y = mx + c represents a linear relationship** (*e.g. plotting logarithmic data from a capacitor discharge experiment to confirm the exponential relationship*)

• **Use of spreadsheet modelling** (*e.g. using a spreadsheet to model how discharge current changes with time*)

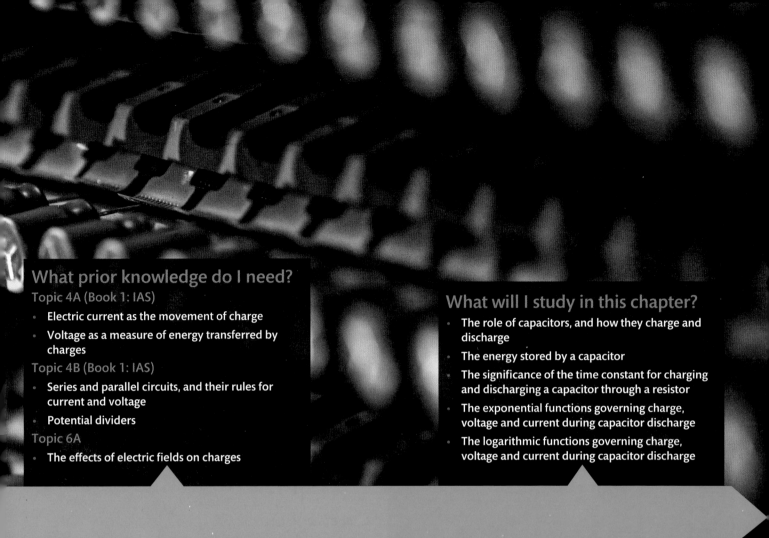

What prior knowledge do I need?

Topic 4A (Book 1: IAS)
- Electric current as the movement of charge
- Voltage as a measure of energy transferred by charges

Topic 4B (Book 1: IAS)
- Series and parallel circuits, and their rules for current and voltage
- Potential dividers

Topic 6A
- The effects of electric fields on charges

What will I study in this chapter?

- The role of capacitors, and how they charge and discharge
- The energy stored by a capacitor
- The significance of the time constant for charging and discharging a capacitor through a resistor
- The exponential functions governing charge, voltage and current during capacitor discharge
- The logarithmic functions governing charge, voltage and current during capacitor discharge

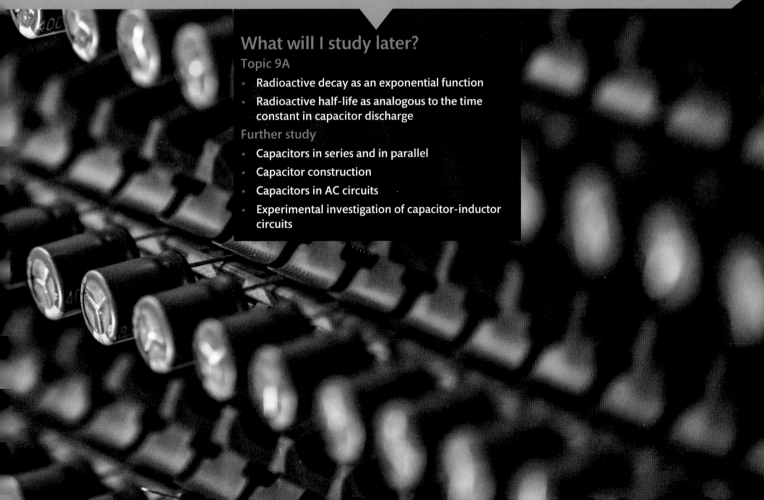

What will I study later?

Topic 9A
- Radioactive decay as an exponential function
- Radioactive half-life as analogous to the time constant in capacitor discharge

Further study
- Capacitors in series and in parallel
- Capacitor construction
- Capacitors in AC circuits
- Experimental investigation of capacitor-inductor circuits

LEARNING OBJECTIVES

- Describe how capacitors can be used in a circuit to store charge.
- Use the equation for capacitance, $C = \dfrac{Q}{V}$.
- Use the equations for energy stored on a capacitor.

STORING CHARGE

We saw in **Section 6A.1** that an electric field can cause charges to move. Indeed, this is why a current flows through a circuit – an electric field is set up within the conducting material and this causes electrons to experience a force and thus move through the wires and components of the circuit. Where there is a gap in a circuit, the effect of the electric field can be experienced by charges across this empty space, but in general, conduction electrons are unable to escape their conductor and move across the empty space. This is why a complete conducting path is needed for a simple circuit to function.

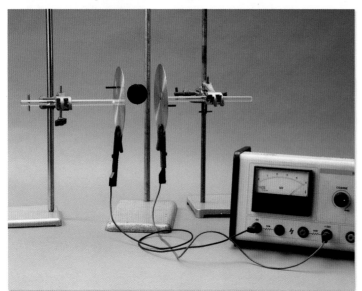

▲ **fig A** An electric field will act across a space. You could test this by hanging a charged sphere near the plates and observing the field's force acting on the sphere.

However, charge can be made to flow in an incomplete circuit. This can be demonstrated by connecting two large metal plates in a circuit with an air gap between them, as in **fig A**. The circuit shown in **fig B** is similar to the situation shown by the photo in **fig A**. When the power supply is connected, the electric field created in the conducting wires causes electrons to flow from the negative terminal towards the positive terminal. Since the electrons cannot cross the gap between the plates, they collect on the plate connected to the negative terminal, which becomes negatively charged. Electrons in the plate connected to the positive terminal flow towards the positive of the cell, which

results in positive charge being left on that plate. The attraction between the opposite charges across the gap creates an electric field between the plates, which increases until the potential difference across the gap is equal to the potential difference of the power supply.

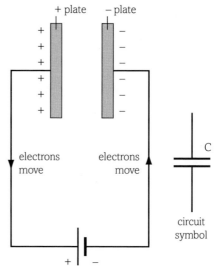

▲ **fig B** A simple capacitor circuit.

A pair of plates such as this with an insulator between them is called a **capacitor**. As we have seen, charge will collect on a capacitor until the potential difference across the plates equals that provided by the power supply to which it is connected. It is then fully charged, and the capacitor is acting as a store of charge. The amount of charge a capacitor can store, per volt applied across it, is called its **capacitance**, C, and is measured in farads (F). The capacitance depends on the size of the plates, their separation and the nature of the insulator between them.

Capacitance can be calculated by the equation:

$$\text{capacitance (F)} = \frac{\text{charge stored (C)}}{\text{potential difference across capacitor (V)}}$$

$$C = \frac{Q}{V}$$

WORKED EXAMPLE 1

(a) What is the capacitance of a capacitor which can store 18 mC of charge when the p.d. across it is 6 V?

$$C = \frac{Q}{V} = \frac{18 \times 10^{-3}}{6} = 3 \times 10^{-3}$$

$$C = 3\,\text{mF}$$

(b) How much charge will be stored on this capacitor if the voltage is increased to 20 V?

$$Q = CV = 3 \times 10^{-3} \times 20 = 60 \times 10^{-3}$$

$$Q = 0.06\,\text{C}$$

Investigating stored charge

A device that will measure the amount of charge directly is called a coulombmeter. By charging a capacitor to various different voltages, and discharging through the coulombmeter each time, you can verify the basic capacitor equation that $C = \dfrac{Q}{V}$. A graph of charge (on the y-axis) against p.d. (on the x-axis) should produce a straight line through the origin. The gradient of this line will equal the capacitance.

▲ **fig C** A coulombmeter will measure how much charge is stored.

ENERGY STORED ON A CHARGED CAPACITOR

A charged capacitor is a store of electrical potential energy. When the capacitor is discharged, this energy can be transferred into other forms. Our definition of voltage gives the energy involved as $E = QV$. However, the energy stored in a charged capacitor is given by $E = \frac{1}{2}QV$. So where has the missing half of the energy gone? This is a trick question, because our original equation assumes that the charge and voltage are constant. However, in order to charge a capacitor, it begins with zero charge stored on it and slowly fills up as the p.d. increases, until the charge at voltage V is given by Q. This can be seen on the graph in **fig D**.

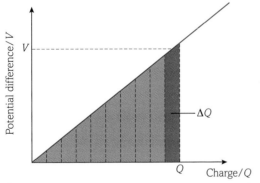

▲ **fig D** Graph of potential difference against charge for a capacitor.

Each time we add a little extra charge (ΔQ) this has to be done by increasing the voltage and pushing the charge on, which takes some energy (we are doing work).

By finding the area of each nearly rectangular strip, we find $V\Delta Q$, which is the amount of extra energy needed for that extra charge. Therefore, the sum of all the strips, or the area under the line, will give us the total energy stored. This is the area of a triangle, so its area is $\frac{1}{2}$ base × height, which from the graph is $\frac{1}{2}QV$.

$$E = \tfrac{1}{2}QV$$

Because $Q = CV$, you can also find two other versions of this equation for the stored energy.

$$E = \tfrac{1}{2}QV = \tfrac{1}{2}(CV)V = \tfrac{1}{2}CV^2$$

Or: $\quad E = \tfrac{1}{2}QV = \tfrac{1}{2}Q\left(\dfrac{Q}{C}\right) = \tfrac{1}{2}\dfrac{Q^2}{C}$

What is the energy stored on a charged $100\,\mu F$ capacitor which has 3 mC of charge?

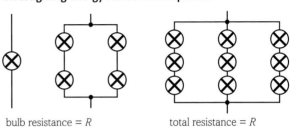

$$E = \tfrac{1}{2}\dfrac{Q^2}{C} = \tfrac{1}{2} \times \dfrac{(3 \times 10^{-3})^2}{(100 \times 10^{-6})}$$

$$E = 0.045\,J$$

Investigating energy stored on a capacitor

bulb resistance = R total resistance = R

▲ **fig E** Investigating how energy stored on a capacitor can be altered.

You can investigate how the energy stored on a capacitor changes with the voltage which is used to charge it. Various combinations of identical series and parallel bulbs will have different overall resistances. If we add an extra parallel branch and increase the number of bulbs on each branch by one, we can keep the total resistance constant, but have more bulbs to light up. The three groups of bulbs in **fig E** all have the same resistance, R. By allowing our charged capacitor to discharge through these different groups of bulbs, and altering the voltage to keep the bulb brightness constant, we can confirm our equation $E = \frac{1}{2}CV^2$ for the energy stored on the capacitor.

> ⚠ Safety Note: High voltage power supplies can give a severe electric shock and should be used with shrouded connectors.

1. What is the capacitance of a capacitor which stores 2 coulombs of charge for every 100 volts applied to it?

2. A 0.01 F capacitor is charged by and then isolated from an 8 V power supply.
   (a) Calculate the charge stored.
   (b) The capacitor is then connected across another identical capacitor, which is uncharged. Describe and explain what will happen to the charge and voltage on each capacitor.

3. How much energy is stored on a $50\,\mu F$ capacitor which is charged to 12 V?

4. A $1200\,\mu F$ capacitor is connected to a voltage supply until fully charged with 10.8 mC. If this capacitor is then disconnected and reconnected across a 10 W light bulb, how long could it light the bulb for?

capacitor an electrical circuit component that stores charge, and so can be used as an energy store

capacitance a measure of the capability of a capacitor; the amount of charge stored per unit voltage across the capacitor, measured in farads, F

LEARNING OBJECTIVES

■ Draw and interpret charge and discharge curves for capacitors.
■ Describe the significance of the time constant, *RC*.

CAPACITOR DISCHARGE CURVES

EXAM HINT

Make sure you know two or three reasons why using a data logger with a voltage sensor is better than having a human taking voltage readings.

PRACTICAL SKILLS   CP11

Investigating current flow through a capacitor

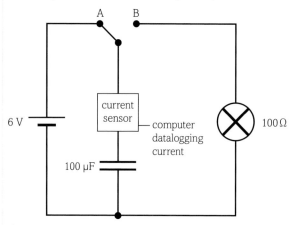

▲ **fig A** Investigating how the current through a capacitor changes over time.

You can investigate how the current through a capacitor changes over time by connecting a datalogger, which senses current in series with the capacitor and then charge and discharge it. A suitable set-up for this is shown in **fig A**. This set-up could be altered to log the potential difference across the capacitor over time, using a voltage sensor in parallel across the capacitor.

 Safety Note: Only use small capacitors as large charged capacitors can give a severe electric shock. Do not overcharge the capacitors as they may explode.

EXAM HINT

Make sure you have a good understanding of this practical as your understanding of the experimental method may be assessed in your exams.

If, in **fig A**, the capacitor is fully charged, it will be at 6 V, and from $Q = CV$ we know it will be storing 0.6 mC of charge. Recall that in **Section 6B.1** we looked at how the electrons in the circuit are influenced by the electric field caused by the supply voltage, and their own mutual repulsion. If the two-way switch in **fig A** is moved to position B, the electrons on the capacitor will be able to move to the positive side of it by discharging via the lamp. As it has 100 Ω resistance, their progress will be slowed by the lamp, but they will still discharge, and the lamp will light for as long as some current flows through it.

At first, the rush of electrons as the capacitor discharges is as high as it can be – the current starts at a maximum. We can calculate this current using Ohm's law – it is 0.06 A. After some electrons have discharged, the p.d. across the capacitor is reduced and the electric field, and therefore the push on the remaining electrons, is weaker. The current $\left(\dfrac{V}{R}\right)$ is less and the light will be dimmer.

Some time later, the flow of electrons is so small that the current is down to a trickle, and the lamp will be so dim that it may look like it is off. Eventually, the capacitor will be fully discharged and there will be no more electrons moving from one side of the capacitor to the other – the current will be zero. If we put this story together over time, the discharging current, p.d. across the capacitor, and charge remaining on the capacitor will follow the patterns shown on the three graphs in **fig B**.

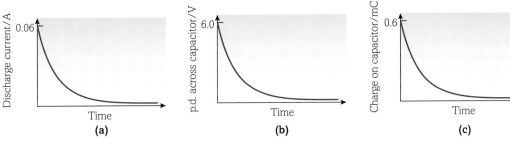

▲ **fig B** Discharge curves for a capacitor through a light bulb.

THE TIME CONSTANT

How could we make the lamp light up for longer, given the same power supply? There are two possibilities:

1 store more charge on the capacitor

2 decrease the rate at which the capacitor discharges.

For the same maximum p.d., increasing the capacitance, C, will increase the charge stored, as $Q = CV$. Alternatively, the charge would flow more slowly if the bulb's resistance, R, was greater.

An overall impression of the rate of discharge of a capacitor can be gained by working out the **time constant**, τ. This is calculated from $\tau = RC$, and with resistance in ohms and capacitance in farads, the answer is in seconds. In fact, the time constant tells you how many seconds it takes for the current to fall to 37% of its starting value. We will see the mathematics of how 37% comes about in the next section, but for now we just need to understand that RC indicates how quickly a charged capacitor will discharge.

WORKED EXAMPLE 2

What is the time constant for the capacitor in the circuit shown in **fig A**?

$\tau = RC = 100 \times 100 \times 10^{-6}$

$\tau = 0.01\,\text{s}$

So the light bulb shown in **fig A** might flash on and off so quickly that we could not see it!

CAR COURTESY LIGHTS

Modern cars often have a courtesy light that comes on when the door is opened, and remains on for a short time after the door is closed. This is useful in case it is dark, allowing the driver to see to put the key in the ignition. The light functions by having a capacitor discharge through the light bulb so that it dims and goes off as the charge runs out. In some cars, the length of time for which the light remains on after the door is closed is adjustable and can be set by the vehicle owner. This adjustable setting makes use of the idea of the time constant, RC. The owner will be able to adjust a switch connected to the courtesy light circuit, which connects more or less resistance to the discharging circuit. So, for the same fully charged capacitor, the time taken to discharge completely will change and the courtesy light illuminates the cabin for more or less time.

▲ **fig C** Capacitor discharge is used in a car courtesy light.

CAPACITOR CHARGING CURVES

By considering the charging process in the same way as we did the discharge of the capacitor in **fig A**, we can quickly work out that the charging process produces graphs such as those in **fig D**.

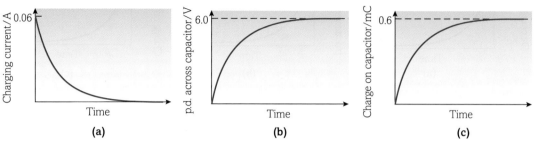

▲ **fig D** Charging curves for a capacitor connected to a 6 V supply.

When charging a capacitor through a resistor, the time constant RC has exactly the same effect. A greater resistance or a larger capacitance, or both, means the circuit will take longer to charge up the capacitor.

CHECKPOINT

SKILLS ANALYSIS, REASONING/ARGUMENTATION

1. ▶ What is the time constant for a car's courtesy light in which a 2 mF capacitor discharges through a 15 kΩ resistor?

2. ▶ Look at the electric circuit of **fig A** and explain the shape of the charging graphs in **fig D** for the situation when the capacitor is initially uncharged and the two-way switch is then connected to point A. In your answer, refer to the movement of electrons within the electric circuit of **fig A**.

3. Draw an accurate sketch graph for the current discharging through a 50 μF capacitor, previously charged by a 6 V supply, and discharged through a 10 kΩ resistance, over a period of 2 seconds.

EXAM HINT

From the time constant equation, if you calculate the capacitance, remember that the unit must be given as F.

SUBJECT VOCABULARY

time constant (for a capacitor resistor circuit) the product of the capacitance and the resistance, giving a measure of the rate for charging and discharging the capacitor. This is the time taken for the current, voltage or charge of a discharging capacitor to fall to 37% of its initial value. Symbol: tau, τ (sometimes T is used instead)

6B | 3 CAPACITOR MATHEMATICS

LEARNING OBJECTIVES

■ Use the equations for exponential discharge in a capacitor resistor circuit.
■ Derive and use capacitor discharge equations in terms of current and voltage, and the corresponding logarithmic equations.

DISCHARGING CAPACITOR MATHS

CHARGE, Q

We have seen that the charging and discharging of a capacitor follows curving graphs in which the current is constantly changing. It follows that the rate of change of charge and p.d. are also constantly changing. These graphs are known as **exponential curves**. The shapes can be produced by plotting mathematical formulae which have power functions in them. In the case of discharging a capacitor, C, through a resistor, R, the function that describes the charge remaining on the capacitor, Q, at a time, t, is:

$$Q = Q_0 e^{-t/RC}$$

where Q_0 is the initial charge on the capacitor at $t = 0$, and e is the special mathematical number which is used in the inverse function of natural logarithms (e $\approx$ 2.718).

WORKED EXAMPLE 1

A 0.03 F capacitor is fully charged by a 12 V supply and is then connected to discharge through a 900 Ω resistor. How much charge remains on the capacitor after 20 seconds?

Initial charge, $Q_0 = CV = 0.03 \times 12 = 0.36\,C$

$Q = Q_0 e^{-t/RC}$

$Q = 0.36 \times e^{(-20/(900 \times 0.03))} = 0.36 \times e^{(-20/27)} = 0.36 \times 0.477$

$Q = 0.17\,C$

VOLTAGE, V

The p.d. across a discharging capacitor will fall as the charge stored falls. By substituting the equation $Q = CV$ into our exponential decay equation, we can show the formula that describes voltage on a discharging capacitor is in exactly the same form as for the charge itself:

$Q = Q_0 e^{-t/RC}$ and $Q = CV$

(which also means that initially, $Q_0 = CV_0$)

$CV = CV_0 e^{-t/RC}$

from which the capacitance term, C, can be cancelled, leaving:

$V = V_0 e^{-t/RC}$

WORKED EXAMPLE 2

A 0.03 F capacitor is fully charged by a 12 V supply and is then connected to discharge through a 900 Ω resistor. What is the p.d. on the capacitor after 20 seconds?

Initial voltage is the same as the supply at 12 V.

$V = V_0 e^{-t/RC}$

$V = 12 \times e^{(-20/(900 \times 0.03))} = 12 \times e^{(-20/27)} = 12 \times 0.477$

$V = 5.7\,V$

CURRENT, I

As we saw in **Section 6B.2**, the discharging current also dies away following an exponential curve. Ohm's law tells us that $V = IR$, and so $V_0 = I_0R$.

$$V = V_0e^{-t/RC}$$
$$IR = I_0Re^{-t/RC}$$

from which the resistance term, R, will cancel on both sides:

$$I = I_0e^{-t/RC}$$

WORKED EXAMPLE 3

A 0.03 F capacitor is fully charged by a 12 V supply and is then connected to discharge through a 900 Ω resistor. What is the discharge current after 20 seconds?

Initial voltage is the same as the supply at 12 V, so the initial current is:

$$I_0 = \frac{V_0}{R} = \frac{12}{900} = 0.013\,A$$

$$I = I_0e^{-t/RC}$$

$$I = 0.013 \times e^{(-20/(900 \times 0.03))} = 0.013 \times e^{(-20/27)} = 0.013 \times 0.477$$

$$I = 6.2\,mA$$

PRACTICAL SKILLS

Using a spreadsheet to investigate the time constant

We can use a spreadsheet to model a timing circuit. This enables us to create a circuit that satisfies the needs of a specific situation (for example, a car courtesy light which needs to stay on for a desired length of time). We can type in different possible values for the circuit components and see what the outcome will be, before building the circuit. **Fig A** shows how such a spreadsheet might appear.

You can create the spreadsheet without doing any experimentation. Give the various cells formulae to calculate what capacitor theory tells us will happen, using the mathematics on these pages. For example, the cell giving the time constant, τ, does not require input from the user – it is programmed to display the multiplication of the capacitance and the discharge resistance. This value is then used in the formula for calculating the values in the current column, using the equation, $I = I_0e^{-t/RC}$, or $I = I_0e^{-t/\tau}$.

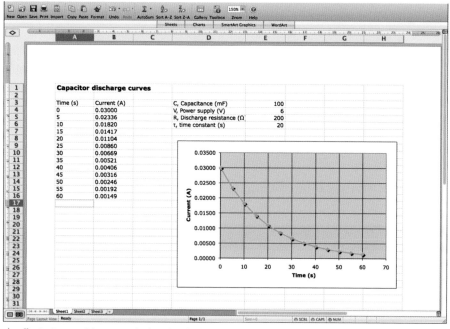

▲ **fig A** A spreadsheet to calculate the discharge curve for a capacitor circuit

CAPACITOR CALCULUS

The mathematics used here, calculus (including integration) is not required in your exam specification in physics. However, for those studying high level mathematics, this will explain the source of the capacitor equations.

The equation for charge on a discharging capacitor is the solution to a differential equation based on considering the rules for voltages around the discharging circuit. With only the capacitor, C, and resistance, R, in the circuit, the e.m.f. is zero. So:

$$0 = V_C + V_R$$

$$V_C = \frac{Q}{C} \text{ and } V_R = IR$$

$$\therefore \quad -IR = \frac{Q}{C}$$

The current is the rate of change of charge.

$$I = \frac{dQ}{dt}$$

So:

$$-\frac{R\,dQ}{dt} = \frac{Q}{C}$$

$$\frac{dQ}{Q} = \frac{-dt}{RC}$$

Integrating this from the start to time t, i.e. from capacitor charge Q_0 to Q:

$$\int_{Q_0}^{Q} \frac{dQ}{Q} = \int_{0}^{t} \frac{-dt}{RC}$$

gives: $\ln Q - \ln Q_0 = \dfrac{-t}{RC}$

or: $\quad \ln Q = \ln Q_0 - \dfrac{t}{RC}$

applying the inverse function of natural logarithm gives:

$$Q = Q_0 e^{-t/RC}$$

LOGARITHMIC EQUATIONS FOR VOLTAGE

We can take the log equation for charge and make substitutions to find equivalent equations for voltage and for current.

$$\ln Q = \ln Q_0 - \frac{t}{RC}$$

$$\therefore \quad \ln(CV) = \ln(CV_0) - \frac{t}{RC}$$

$$\therefore \quad \ln C + \ln V = \ln C + \ln V_0 - \frac{t}{RC}$$

C is fixed, so $\ln C$ is the same on both sides.

$$\therefore \quad \ln V = \ln V_0 - \frac{t}{RC}$$

Applying the inverse function of natural logarithm gives:

$$V = V_0 e^{-t/RC}$$

LOGARITHMIC EQUATIONS FOR CURRENT

Similarly, we can take the log equation for voltage and make substitutions to find equivalent equations for current.

$$\ln V = \ln V_0 - \frac{t}{RC}$$

$$\therefore \quad \ln(IR) = \ln(I_0 R) - \frac{t}{RC}$$

$$\therefore \quad \ln I + \ln R = \ln I_0 + \ln R - \frac{t}{RC}$$

R is fixed, so $\ln R$ is the same on both sides.

$$\therefore \quad \ln I = \ln I_0 - \frac{t}{RC}$$

Applying the inverse function of natural logarithm gives:

$$I = I_0 e^{-t/RC}$$

THE '37% LIFE'

If we consider the charge at time τ:

$$t = \tau = RC \text{ so } -\frac{t}{RC} = -\frac{RC}{RC} = -1$$

$$\therefore \quad Q = Q_0 e^{-RC/RC}$$

$$Q = Q_0 e^{-1} \text{ and } e^{-1} = 0.37$$

So: $\quad Q = 0.37 Q_0$

The charge is 37% of its original value.

This shows that the time constant describes the decay of charge on a discharging capacitor in exactly the same way as radioactive half-life describes the number of radioactive nuclei remaining (see **Section 9A.2**), except that instead of describing the time taken to reach half of the initial value, τ is the time taken to reach 37% of the initial value. This similarity comes from the fact that radioactive decay also follows an exponential equation: $N = N_0 e^{-\lambda t}$.

SKILLS ADAPTABILITY

1. Use the data in **fig A** to produce a spreadsheet of data and plot a graph of current discharge over time for this capacitor and resistor.
   (a) Use your graph to find the current through the capacitor after 30 s.
   (b) If this were a model of an automatic hand dryer circuit that requires 4.0 V to operate, use your graph to work out how long it will remain on.
   (c) What would you change so that the dryer remains on for 30 s?

2. A 200 mF capacitor is charged to 8 V. If it is then discharged through a 4.7 Ω resistor, what would the discharge current be after 3.5 s?

SKILLS DECISION MAKING, INTERPRETATION

3. $\ln V = \ln V_0 - \dfrac{t}{RC}$

   An experiment is conducted to find the value of an unknown capacitance discharging through a known resistor. Consider the equation $y = mx + c$ and explain why the log version of the capacitor voltage equation might be more useful than the exponential version of the equation.

EXAM HINT

When reading from a graph, always check the values above and below the point you are looking at to be sure you apply the correct scale.

SUBJECT VOCABULARY

exponential curves mathematical functions generated by each value being proportional to the value of one variable as the index of a fixed base: $f(x) = b^x$

ULTRACAPACITORS

Generally, we have not made the switch from petrol cars to electric cars very quickly. Most people think that they are not able to store enough energy, they have a limited range between charges and they take a long time to charge. Ultracapacitors have been suggested as a possible solution to these problems, as a replacement for electrochemical rechargeable batteries.

PAPER FOR SCIENTIFIC JOURNAL

PAPER-BASED ULTRACAPACITORS WITH CARBON NANOTUBES–GRAPHENE COMPOSITES

▲ **fig A** Electron microscope image of graphene flakes showing exceptionally high surface area, due to extreme convolutions.

In this paper, paper-based ultracapacitors were fabricated by the rod-rolling method with the ink of carbon nanomaterials, which were synthesized by arc discharge under various magnetic conditions. Composites of carbon nanostructures, including high-purity single-walled carbon nanotubes (SWCNTs) and graphene flakes, were synthesized simultaneously in a magnetically enhanced arc. These two nanostructures have promising electrical properties and synergistic effects in the application of ultracapacitors. Scanning electron microscope, transmission electron microscope, and Raman spectroscopy were employed to characterize the properties of carbon nanostructures and their thin films. The sheet resistance of the SWCNT and composite thin films was also evaluated by four-point probe from room temperature to the cryogenic temperature as low as 90 K. In addition, measurements of cyclic voltammetry and galvanostatic charging/discharging showed the ultracapacitor based on composites possessed a superior specific capacitance of up to 100 F/g, which is around three times higher than the ultracapacitor entirely fabricated with SWCNT.

SCIENCE COMMUNICATION

The extract opposite consists of information from a peer-reviewed scientific paper.

1 (a) Discuss the tone and level of vocabulary included in the article. Who is the intended audience?

(b) Discuss the level of scientific detail included in the extract, particularly considering the intended audience.

INTERPRETATION NOTE

If you search online for 'graphene ultracapacitors' there are several articles in more everyday language that explain the main points of this scientific paper, for comparison.

PHYSICS IN DETAIL

Now we will look at the physics in detail. Some of these questions will link to topics elsewhere in this book, so you may need to combine concepts from different areas of physics to work out the answers.

2 (a) Explain the meaning of the phrase 'specific capacitance' considering that it was quoted as being 100 F g^{-1}.

(b) How does the specific capacitance affect the potential use of a capacitor as an energy store in an electric car?

3 Explain the importance of the resistance of a capacitor.

4 The fundamental purpose of a capacitor is to store charge. Graphene flakes have an exceptionally high surface area as shown in **fig A**. Explain why the incorporation of graphene flakes into capacitor construction could make very high capacitances possible.

PHYSICS TIP

Consider how we calculate the time constant, τ, as well as the idea of internal resistance.

ACTIVITY

Draw a flowchart detailing the steps that the authors say their paper will explain to a reader. Your flowchart can use technical terms, but should be written in everyday language as much as is possible.

THINKING BIGGER TIP

Your flowchart should be in the chronological order that the scientists would have had to work on building and testing the materials and then building their ultracapacitors.

DID YOU KNOW?

Isolated capacitors can hold their charge for years without losing much at all. It is important that any unfamiliar large capacitor is assumed to be fully charged, as it could give a shock if it discharges through you.

6B EXAM PRACTICE

[Note: In questions marked with an asterisk (*), marks will be awarded for your ability to structure your answer logically, showing how the points that you make are related or follow on from each other.]

1 A correct unit for the time constant in a capacitor discharge circuit is:

A $\Omega\,m$

B s^{-1}

C s

D F [1]

(Total for Question 1 = 1 mark)

2 A 100 mF capacitor is charged by connecting to a 1.5 V cell and it stores 0.1125 J of energy. If the cell is replaced by a 3.0 V battery:

A the charge stored halves and the energy stored increases to 0.225 J

B the charge stored halves and the energy stored increases to 0.45 J

C the charge stored doubles and the energy stored increases to 0.225 J

D the charge stored doubles and the energy stored increases to 0.45 J. [1]

(Total for Question 2 = 1 mark)

3 Which of the following equations could correctly connect the discharge current, I, from a capacitor, C, through a resistor, R, at a time, t, if the initial discharge current was I_0?

A $I = \ln I_0 - \dfrac{t}{RC}$

B $I = I_0 - \dfrac{t}{RC}$

C $\ln I = \ln I_0 - \dfrac{t}{RC}$

D $\ln I = \ln I_0 - \ln\dfrac{t}{RC}$ [1]

(Total for Question 3 = 1 mark)

4 Which row in the table correctly shows how the charge on and voltage across a capacitor change over time as the capacitor discharges?

	Charge	Voltage
A	Increases exponentially	Increases exponentially
B	Increases exponentially	Decreases exponentially
C	Decreases exponentially	Increases exponentially
D	Decreases exponentially	Decreases exponentially

[1]

(Total for Question 4 = 1 mark)

5 The diagram shows a circuit that includes a capacitor.

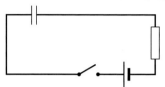

(a) (i) Explain what happens to the capacitor when the switch is closed. [2]

(ii) The potential difference (p.d.) across the resistor rises to a maximum as the switch is closed.
Explain why this p.d. subsequently decreases to zero. [2]

*(b) One type of microphone uses a capacitor. The capacitor consists of a flexible front plate (diaphragm) and a fixed back plate. The output signal is the potential difference across the resistor.

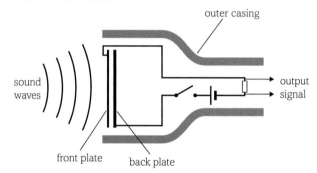

The sound waves cause the flexible front plate to vibrate and change the capacitance. Moving the plates closer together increases the capacitance. Moving the plates further apart decreases the capacitance.

Explain how the sound wave produces an alternating output signal. [4]

(c) A microphone has a capacitor of capacitance 500 pF and a resistor of resistance 10 MΩ.

Explain why these values are suitable even for sounds of the lowest audible frequency of about 20 Hz. [4]

(Total for Question 5 = 12 marks)

6 A student needs to order a capacitor for a project. He sees this picture on a website accompanied by this information: capacitance tolerance ±20%.

16V
10 000μF

16V
10 000μF

Taking the tolerance into account, calculate

(a) the maximum charge a capacitor of this type can hold   [3]

(b) the maximum energy it can store.   [2]

(Total for Question 6 = 5 marks)

7 A designer needs a circuit that will cause a delay in switching off the interior light in a car after the door is shut.

She uses a circuit with a resistor and a capacitor. She knows that the time constant T is given by:

$$T = RC$$

where R is the resistance in ohms and C is the capacitance in farads.

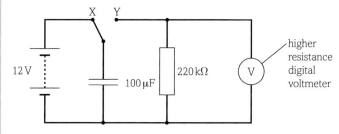

X   Y

12 V

100 μF   220 kΩ   V   higher resistance digital voltmeter

With the switch in position X, the capacitor is charged to 12 V. When the switch is moved to position Y, the capacitor discharges through the resistor and the potential difference (p.d.) across the resistor falls steadily from 12 V.

(a) (i) Calculate a theoretical value for the time constant for this circuit.   [1]

    (ii) What is the significance of the time constant for such a discharge?   [1]

(b) The designer decides to check the theoretical value for the time constant T using a stopwatch, with a precision of 0.01 s.

    (i) State why the voltmeter needs to have a high resistance.   [1]

    (ii) State why a stopwatch is suitable for measuring the time in this context.   [1]

    (iii) State what she should do to make her value for T as reliable as possible.   [1]

(c) For a capacitor discharging through a resistor, the potential difference V across the resistor at time t is given by

$$V = V_0\, e^{-t/RC}$$

Explain why a graph of $\ln V$ against t should be a straight line.   [2]

(d) The designer uses the circuit to obtain the following data.

t/s	V/V	
0	12.00	
5	9.41	
10	7.16	
15	5.49	
20	4.55	
25	3.49	
30	2.68	
35	2.04	

Plot a graph to show that these data are consistent with $V = V_0\, e^{-t/RC}$.

Use the extra column in a table like that above for your processed data.   [4]

(e) (i) Use your graph to determine another value for time constant.   [2]

    (ii) Calculate the percentage difference between your value from the graph and the theoretical value from (a) (i).   [1]

(f) (i) Use your graph to determine how long it takes for the p.d. to decrease to 5.0 V.
Add to your graph to show how you did this.   [2]

    (ii) The designer wants the p.d. to decrease to 5 V in about 12 s.
Choose the value of R she should use.   [1]

    A 47 kΩ

    B 100 kΩ

    C 150 kΩ

    D 330 kΩ

(Total for Question 7 = 17 marks)

TOPIC 6 ELECTRIC AND MAGNETIC FIELDS

CHAPTER **6C**

ELECTROMAGNETIC EFFECTS

Industrial electricity generation is vital in our technological world. Nearly every device is electrically powered now, so the physics behind generating electricity is very important. Electric and magnetic fields are connected; changing one either automatically generates or changes the other.

As with all things, the conservation of energy is very important and this affects the change of movement into electricity. However, electrical energy being transferred to kinetic energy is also very important in today's world. This is the opposite of the generation of electricity, and we will look at it in this chapter too. Other machines such as particle accelerators and mass spectrometers are also affected by the equations in this chapter.

Finally, we will see how these effects make electrical transformers work. The way transformers change energy from electric current into magnetic fields and back again is the closest people have come to creating a perfectly efficient machine.

MATHS SKILLS FOR THIS CHAPTER

- **Recognising and use of appropriate units** (*e.g. the tesla*)
- **Substituting numerical values into equations** (*e.g. calculate magnetic flux linkage*)
- **Understanding the slope of a tangent to a curve as a measure of the rate of change** (*e.g. induced e.m.f. as the rate of change of magnetic flux linkage*)
- **Use of angles in 2D and 3D structures** (*e.g. finding the magnetic flux linkage when a coil area is at an angle to the magnetic field direction*)
- **Distinguishing between instantaneous and average rates of change** (*e.g. comparing the e.m.f. induced at different times as a magnet falls through a coil*)

What prior knowledge do I need?

- Magnetic field rules for attraction and repulsion

Topic 1A (Book 1: IAS)

- Resolving vectors
- Conservation of energy

Topic 4A (Book 1: IAS)

- Electric current as the movement of charge

Topic 5B

- The centripetal force equation

Topic 6A

- The effects of electric fields on charges

What will I study in this chapter?

- How to explain the strength of a magnetic field
- The effects of electric and magnetic fields on the movement of charged particles
- How electricity can be generated
- How to calculate the amount and direction of induced e.m.f.
- How electrical transformers work

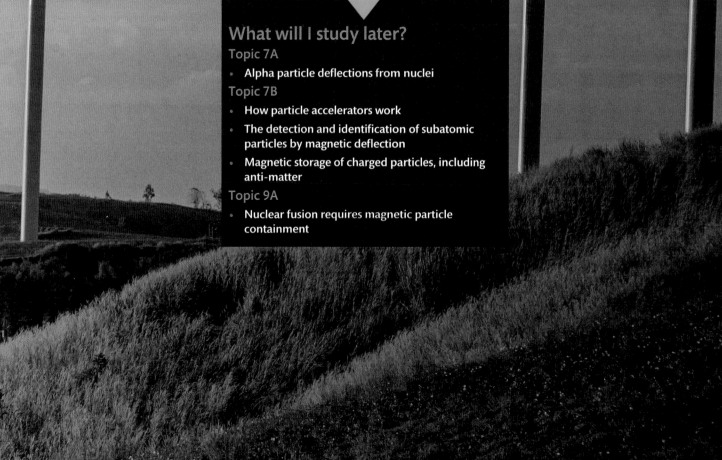

What will I study later?

Topic 7A

- Alpha particle deflections from nuclei

Topic 7B

- How particle accelerators work
- The detection and identification of subatomic particles by magnetic deflection
- Magnetic storage of charged particles, including anti-matter

Topic 9A

- Nuclear fusion requires magnetic particle containment

6C 1 MAGNETIC FIELDS

LEARNING OBJECTIVES

■ Define the terms magnetic flux density, B, magnetic flux, Φ, and flux linkage, $N\Phi$.

■ Calculate flux, flux density and flux linkage.

MAGNETIC FLUX

In **Section 6A.1** we saw how an electric field can be represented by lines that show how a charge will experience a force when placed in the field. You are probably more familiar with the everyday effects of magnetic fields such as attracting a fridge with a magnetic souvenir. Usually the fridge magnet moves towards the fridge rather than the other way around, but Newton's third and second laws of motion explain this outcome. Electric and magnetic fields are very similar (and utterly intertwined) as we saw in **Book 1, Topic 3** with the nature of electromagnetic radiation.

Where an electric field affects charges, a magnetic field affects **poles**. A place that will cause a magnetic pole to experience a force is called a **magnetic field**.

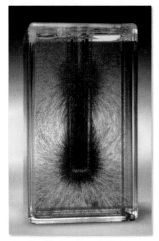

▲ **fig A** A magnetic field around a permanent bar magnet.

When we draw a magnetic field in a diagram, the field lines show the direction in which a north pole will be pushed. Magnetic poles always exist in north and south pairs, but we take the field as acting from north to south, and this is the direction of arrows drawn onto magnetic field lines (also called lines of **magnetic flux**). Like electric field patterns, the closer the lines are together, the stronger the field is. The term referring to the strength of a magnetic field is the **magnetic flux density**, B, and the SI unit for it is the **tesla (T)**.

Looking at **fig B**, you can see that the field lines all go in and out of the ends of the bar magnet. This forces them to be closer together at those points, which means that the field is stronger

there. This is why a paperclip picked up by a bar magnet will jump to one of its ends – as the attraction is stronger at the end. The quantity of flux, Φ (measured in **weber, Wb**), through any given area indicates the strength of the effect of the field there. This can be calculated for a particular region by multiplying the area which is enclosed by the region by the component of flux density perpendicular to the area, A.

$$\Phi = B\sin\theta \times A$$

If the magnetic flux is in a uniform direction and lies perpendicular to A, then $\sin 90° = 1$, and the equation simplifies to:

$$\Phi = BA$$

WORKED EXAMPLE 1

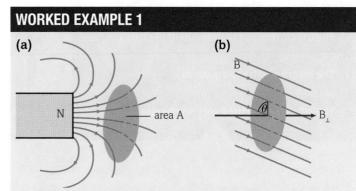

▲ **fig B** Magnetic flux contained in a small area, A.

(a) If the bar magnet in **fig B(a)** causes a magnetic field with a flux density of 20 mT perpendicular to area A, how much flux will be contained by this area if it is 10 cm²?

$$A = 10\,cm^2 = 10 \times 10^{-4}\,m^2$$

$$\Phi = B \times A = 20 \times 10^{-3} \times 10 \times 10^{-4}$$

$$\Phi = 2.0 \times 10^{-5}\,Wb$$

(b) In **fig B(b)**, the uniform magnetic field has a flux density of 5×10^{-5} T and is at an angle of 75° to the region of area A. How much flux will be contained by this area if it is 1 cm²?

$$A = 1\,cm^2 = 1 \times 10^{-4}\,m^2$$

$$\Phi = B\sin\theta \times A = 5 \times 10^{-5} \times \sin 75° \times 1 \times 10^{-4}$$

$$\Phi = 4.8 \times 10^{-9}\,Wb$$

MAGNETIC FLUX DENSITY

In the situation where the flux is perpendicular to the area, a rearrangement of the equation shows why the quantity B is known as the magnetic flux *density*:

$$B = \frac{\Phi}{A}$$

By sharing the flux over the area, B indicates how close together the magnetic field lines are – how dense they are. How dense field lines are indicates how strong the field is, so B is also known as the **magnetic field strength**.

FLUX LINKAGE

The interaction between magnetic fields and charged particles, or conductors, allows motors to operate, and electricity to be generated. In most practical applications, magnetic flux is made to interact with a coil of wire, as the effect on a single strand of wire is too small to be useful. If the single wire is coiled up, then the magnetic field can interact with each turn on the coil, and so any effect is multiplied N times, where N is the number of turns on the coil.

The amount of magnetic flux interacting with a coil of wire is known as the **flux linkage**, and is the product of the number of turns of wire and the flux in that region:

flux linkage = $N\Phi$ measured in **weber-turns**

Remembering that $\Phi = BA$ means that we also have:

flux linkage = BAN

WORKED EXAMPLE 2

Zoran takes a wire and winds it into a coil of ten circular turns with a radius of 5 cm and then holds this in a magnetic field with a strength of 20 mT. What is the flux linkage?

area, $A = \pi \times (0.05)^2 = 7.85 \times 10^{-3}$ m^2

flux linkage = BAN = 0.02 × (7.85 × 10^{-3}) × 10

flux linkage = 1.6 × 10^{-3} Wb

CHECKPOINT

1. The Earth acts as a giant bar magnet, and the geographic North Pole is at the magnetic south end of the field.
   (a) Draw a sketch of the Earth with its magnetic field.
   (b) Explain where you would expect the Earth's magnetic flux density to be the greatest.

2. ▶ In Britain, the Earth's magnetic flux density is 50 000 nT, with the field close to horizontal. Imagine you are in Britain.
   (a) Estimate the flux passing through your body if you stand vertically and face north.
   (b) A typical smartphone compass uses a small magnetometer chip to determine which way the phone is facing. Such chips are typically 2 mm by 2 mm squares. If you held a smartphone so the magnetometer chip was at an angle of 20° to the horizontal, what would be the flux passing through the chip for it to measure?

3. In an electric motor, a permanent magnet of flux density 18.4 mT has its field placed perpendicular to a coil of 250 turns of copper wire. The diameter of the coil is 8.50 cm. What is the flux linkage between the field and the coil?

SKILLS   PROBLEM-SOLVING

SUBJECT VOCABULARY

poles the magnetic equivalent of a charge on a particle: north pole or south pole

magnetic field a region of space that will cause a magnetic pole to feel a force

magnetic flux an alternative phrase referring to magnetic field lines

magnetic flux density the ratio of magnetic flux to the area it is passing through

tesla (T) the unit for magnetic flux density, or magnetic field strength

weber, Wb the unit of measurement of magnetic flux, Φ, (and magnetic flux linkage, $N\Phi$)

magnetic field strength an alternative phrase for magnetic flux density

flux linkage the amount of magnetic flux interacting with a coil of wire

weber-turns the unit for magnetic flux linkage

LEARNING OBJECTIVES

■ Apply Fleming's left hand rule to current-carrying conductors in a magnetic field.

FLEMING'S LEFT HAND RULE

Magnetic fields can affect moving electric charges, as well as magnetic poles. If you place a wire in a magnetic field and pass a current through it, the wire will experience a force on it (**fig A**). This is called the **motor effect**. The effect is greatest when the wire and the magnetic field are at right angles. In this instance, the force will be at right angles to both, in the third dimension, as shown by **Fleming's left hand rule** in **fig B**.

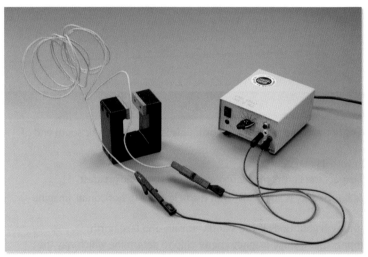

▲ **fig A** The jumping wire experiment illustrates the motor effect in action.

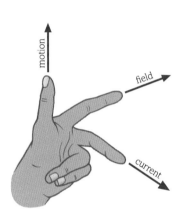

▲ **fig B** Fleming's left hand rule gives the relative directions of the field, current and movement in the motor effect.

WORKED EXAMPLE 1

Look at the diagram in **fig C** in which a wire connected to a cell is placed in a magnetic field. In what direction will the wire experience a force?

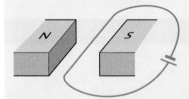

▲ **fig C** Fleming's left hand rule in action.

The conventional current will flow from the positive end of the cell, so within the magnetic field, as observed in **fig D**, it will be moving away from us. With the magnetic field from north to south (towards the right in **fig D**), Fleming's left hand rule tells us that the wire will experience a force pushing it downwards.

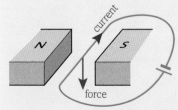

▲ **fig D** This wire will feel a downwards force.

LEARNING TIP

If a current is parallel to a magnetic field, there is no force.

PRACTICAL SKILLS

Build your own motor

The forced movement of a current-carrying conductor within a magnetic field is the principle that causes motors to work. From Fleming's left hand rule, it is clear that a coil of wire experiences a turning force if the wire is carrying a current and is put in a magnetic field. This is because the current travels in opposite directions on opposite sides of the coil which causes forces in opposite directions and rotates the coil. If it is free to move, then the coil (or motor) will spin continuously. If you use the apparatus in **fig E**, remember to use a low power d.c. supply.

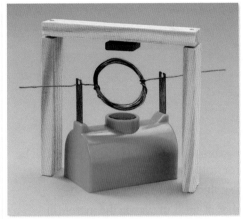

▲ **fig E** Motors obey Fleming's left hand rule.

 Safety Note: Use only a low voltage battery or d.c. power supply to power the electric motor.

THE TINIEST MOTOR IN THE WORLD?

There was some amazement amongst scientists when the University of Berkeley in California produced an electric motor which was not much more than 100 µm across. However, this 'micromotor' now seems like a giant, when compared with the Berkeley lab's 'nanomotor'. In 2003, the same Zettl Lab at Berkeley produced a motor which was less than 500 nanometres across. In 2014, the Cockrell School of Engineering at the University of Texas built a motor of similar tiny size, but rotated it faster and for longer than any previous nanomotor.

▲ **fig F** Computer generated image of a nanomotor: a tiny gold rotor spins on a carbon nanotube axle. The entire set-up would fit within one wavelength of red light.

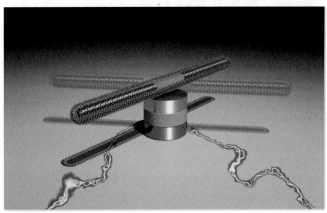

▲ **fig G** The tiniest motor possible? Built by researchers at the Cockrell School of Engineering at The University of Texas, this was the smallest, fastest and longest-running tiny synthetic motor in 2014.

Moving systems on a molecular scale have also been produced by other groups of scientists using chemical systems to generate the forces, but the motors shown in **fig F** and **fig G** are the smallest to use electromagnetic forces as explained in this section.

CHECKPOINT

SKILLS ⮞ COMMUNICATION

1. ⮞ Describe the use of Fleming's left hand rule.
2. Copy the diagrams in **fig H** and draw an arrow to show the direction of any force acting on the wire in each case.

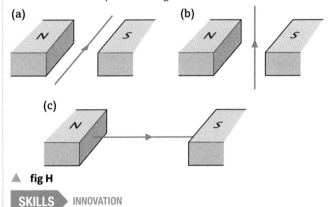

▲ **fig H**

SKILLS ⮞ INNOVATION

3. ⮞ Many real motors use electromagnets to create the magnetic field that causes their rotor to spin. Give one advantage and one disadvantage of building a motor which uses electromagnets rather than permanent magnets.

EXAM HINT

If a question asks you to **Describe** something, your description may have linked points, but does not need to include any justification or reasons.

SUBJECT VOCABULARY

motor effect a wire carrying a current, held within a magnetic field, will experience a force
Fleming's left hand rule a rule for determining the direction of the force generated by the motor effect

LEARNING OBJECTIVES

- Use the equation $F = BIL \sin\theta$, for a current-carrying conductor in a magnetic field.
- Apply Fleming's left hand rule to charged particles moving in a magnetic field.
- Use the equation $F = Bqv \sin\theta$, for a charged particle moving in a magnetic field.

FORCES ON WIRES: $F = BIL$

PRACTICAL SKILLS

Investigating the magnetic force on a current-carrying conductor
You can investigate how much force will be on a wire experiencing the motor effect. In the experiment shown in **fig A**, it is easy to alter the current through the wire, the length that is within the magnetic field, and the angle it cuts across the field.

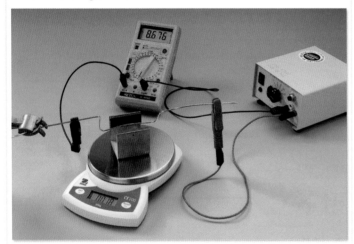

▲ **fig A** Investigating the strength of the force on a current-carrying conductor in a magnetic field.

⚠ Safety Note: The wire will get hot enough to burn skin if left connected for too long.

The strength of the force, F, on a wire which has a current, I, through it whilst it is in a magnetic field, B, is given by the equation:

$F = B \times I \times L \times \sin\theta$

where L is the length of the wire within the field, and θ is the angle the current makes with the lines of the magnetic field. For simplicity, we will only consider uniform magnetic fields in which the field lines are all parallel.

It is common to set up situations in which the angle θ is 90° so that $\sin\theta$ is a maximum and equals 1. This reduces the formula to $F = BIL$.

WORKED EXAMPLE 1

Hanul sets up a jumping wire demonstration. He uses a wire with a current of 2 A running through it, and a pair of magnets that have a magnetic field of 0.5 mT. He is not too careful in setting up and 5 cm of the wire actually hangs across the field at an angle of 80°. How much force does Hanul's wire experience? And, if it has a mass of 9 grams, how fast will it initially accelerate?

$$F = BIL \sin\theta = 0.5 \times 10^{-3} \times 2 \times 0.05 \times \sin 80°$$

$$F = 4.92 \times 10^{-5}\,\text{N}$$

$$a = \frac{F}{m} = \frac{4.92 \times 10^{-5}}{0.009}$$

$$a = 5.5 \times 10^{-3}\,\text{m s}^{-2}$$

A consequence of the expression $F = BIL$ is that a motor can be made more powerful, or faster, by:

- increasing the current through the motor (I)
- increasing the number of turns of wire in the motor (L)
- increasing the magnetic field within the motor (B).

The magnetic field strength is usually maximised by making the coil's core out of soft iron. Some motors use electromagnets to provide the field, and these could be strengthened by increasing the current through them.

FORCES ON PARTICLES: $F = Bqv$

The motor effect happens because a charged particle moving at right angles to a magnetic field experiences a force on it at right angles to its direction of motion and also at right angles to the field. If the charged particle is constrained – like an electron in a current in a wire – then the force will transfer to the wire itself. If the particle is flying freely, its direction will change and it will travel a circular path whilst in the magnetic field (**fig B**). The direction of the force acting on a charged particle moving in a magnetic field can be found using Fleming's left hand rule. We must be careful with the current direction though: a negative charge moving in one direction is a current flowing in the opposite direction.

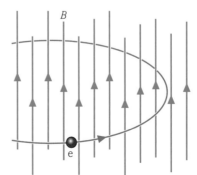

▲ **fig B** Charged particles moving in a magnetic field follow a circular path as the motor effect provides a centripetal force.

The strength of the force on a charged particle moving across a magnetic field is given by the equation:

$$F = B \times q \times v \times \sin\theta$$

where q is the charge on the particle, v is its velocity, and θ is the angle between the velocity and the magnetic field lines. A simplified situation that is often considered is that of an electron (charge 'e') moving at right angles to the field ($\sin 90° = 1$). This reduces the formula to $F = Bev$.

PRACTICAL SKILLS

Investigating the magnetic force on a beam of charged particles

▲ **fig C** Investigating the force on charged particles moving in a magnetic field.

You can investigate how much force will be on an electron moving across a magnetic field using the equipment shown in **fig C**. The grid allows you to observe the path travelled by the electrons, and there is a standard formula for calculating the magnetic field strength provided by a pair of parallel electromagnetic coils, often referred to as Helmholtz coils.

Safety Note: The power supplies can give a severe electric shock and should be used with shrouded connectors. Do not remove or attach connectors with the power switched on.

As the force on the charged particle is always at right angles to the direction of its velocity, it acts as a centripetal force, and the particle follows a circular path. As the charged particle's velocity is constantly changing along a circular path, it is being accelerated by the magnetic field. This means that given the right combination of conditions, a moving charged particle could be held by a magnetic field, continuously orbiting a central point. This is the principle by which artificially generated anti-matter is contained, to save it from annihilation, for future use or study.

THE MASS SPECTROMETER

EXAM HINT: EXTRA CONTENT

The content in this section has been included to illustrate how different areas of physics can be brought together in one experiment to solve a problem. It is not required by the specification.

Scientists often need to identify unknown chemicals. This is particularly important, for example, in the field of forensic science, where a crime scene technician will take a sample of unknown material which they need to identify. A machine called a mass spectrometer (**fig D**) can separate chemicals according to their charge/mass ratio, which allows unique identification of each substance within a sample.

A chemical to be identified enters the machine and is ionised. This charge will then allow it to be accelerated in two different ways within the mass spectrometer. An electric field increases its speed. Then it experiences a force when it travels through the field of the electromagnet, which changes its direction.

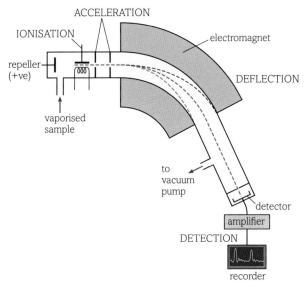

▲ **fig D** Schematic of the basic parts of a mass spectrometer.

The force on a charged particle moving at right angles to a magnetic field is given by:

$$F = Bqv$$

Centripetal force is given by:

$$F = \frac{mv^2}{r}$$

So for particles following the circular curve through the electromagnet on a path to the detector:

$$Bqv = \frac{mv^2}{r}$$

which rearranges to: $\dfrac{q}{m} = \dfrac{v}{Br}$

The charge/mass ratio will identify the particles involved, so with B and r (the radius of the curvature of the particle through the mass spectrometer) known from the calibration of the machine, all we need to know is how fast the particles were moving when they entered the electromagnet. The electric field acts on their charge and accelerates them to this speed, and the kinetic energy gained comes from the potential difference, V, that they pass through according to:

$$\frac{1}{2}mv^2 = qV$$

which gives: $v = \sqrt{\dfrac{2qV}{m}}$

Substituting this into our equation for the charge/mass ratio:

$$\frac{q}{m} = \frac{\sqrt{2qV/m}}{Br}$$

Squaring gives:

$$\frac{q^2}{m^2} = \frac{2qV/m}{B^2r^2}$$

$$\frac{m}{q} \times \frac{q^2}{m^2} = \frac{m}{q} \times \frac{2qV}{mB^2r^2}$$

$$\frac{q}{m} = \frac{2V}{B^2r^2}$$

By changing the accelerating voltage and the strength of the electromagnet (by changing the current through it) we can identify the various chemicals contained within a sample, as each is measured by the detector. The intensity of the current in the detector can indicate the proportion of a given chemical within the sample.

WORKED EXAMPLE 2

A company selling sea salt for cooking wanted to find out if its ratio of chlorine isotopes was different from the usual ratio. In nature, the ^{35}Cl isotope is more common than the ^{37}Cl isotope by a factor of $3:1$. The company analysed a sample of their product using a mass spectrometer that would ionise the chlorine atoms into single minus ions. What would be the difference between the radii of curvature caused for the chlorine isotopes, if the mass spectrometer operated at an accelerating potential of 3.0 kV and had a magnetic field strength of 3.0 T?

$$\frac{q}{m} = \frac{2V}{B^2r^2}$$

$$r^2 = \frac{2Vm}{B^2q}$$

$$r = \sqrt{\frac{2Vm}{B^2q}}$$

For $^{35}Cl^-$:

$$r = \sqrt{\frac{2 \times 3000 \times (35 \times 1.67 \times 10^{-27})}{(3.0)^2 \times 1.6 \times 10^{-19}}}$$

$$r = 0.0156\,m$$

For $^{37}Cl^-$:

$$r = \sqrt{\frac{2 \times 3000 \times (37 \times 1.67 \times 10^{-27})}{(3.0)^2 \times 1.6 \times 10^{-19}}}$$

$$r = 0.0160\,m$$

Therefore the difference in the radius of curvature of the path of these two ions is 0.4 mm.

EXAM HINT

If a question (as in question 1(c)) asks you to **Comment on** X, you will need to connect together at least two pieces of information to give a judgement about X.

SKILLS ADAPTIVE LEARNING

SKILLS REASONING/ARGUMENTATION

CHECKPOINT

1. (a) How much force would be experienced by a 12 cm wire carrying 0.8 A perpendicular to the Earth's surface magnetic field of 5×10^{-5} T?
    (b) How much force would be experienced by a proton travelling across the Earth's magnetic field at $500\,m\,s^{-1}$?
    (c) How fast would a proton need to travel in order for the electromagnetic force on it to make it orbit the Earth at the surface? (Radius of Earth = 6.4×10^6 m.) Comment on your answer.

2. An electron beam travels across a 36 mT magnetic field at 45° to the field. If the electrons are travelling at 10% of the speed of light, what force does the field exert on each electron?

3. ▶ Speed = $\frac{distance}{time}$ and current = $\frac{charge}{time}$. Explain how $F = BIL$ is actually the same equation as $F = Bqv$ but considered for many charges in a group.

4. ▶ For the investigation on salt in the worked example on this page, calculate the difference in the radii of curvature that would be found if the company investigated the two isotopes of sodium: $^{23}Na^+$ and $^{22}Na^+$. Explain why such small differences can be easily detected by a machine like that shown in **fig D**.

LEARNING OBJECTIVES

■ Explain the factors affecting the e.m.f. induced in a coil when there is relative motion between the coil and a permanent magnet.

■ Explain the factors affecting the e.m.f. induced in a coil when there is a change of current in another coil linked with this coil.

■ Define Faraday's law, and be able to use the equation

$$\varepsilon = -\frac{d(N\Phi)}{dt}.$$

ELECTROMAGNETIC INDUCTION

We have seen that the movement of a charged particle in a magnetic field causes it to experience a force. Newton's third law of motion reminds us that this force must have a counterpart that acts equally in the opposite direction. This pair of electromagnetic forces is generated whenever there is relative motion between a charge and a magnetic field. So, a magnetic field moving past a stationary charge will create the same force. The velocity term in the expression $F = Bqv$ actually refers to the relative (perpendicular) velocity between the magnetic field lines and q. (Also remember that if the movement is not at right angles, then we need to work out the component of it that is at right angles by including the $\sin \theta$ term: $F = Bqv\sin \theta$.)

This means that if we move a magnet near a wire, the electrons in the wire will experience a force to make them move through the wire. This is an e.m.f.; if the wire is in a complete circuit, then the electrons will move, forming an electric current. We can use this principle to generate electricity. Reversing the direction of the magnetic field, or the direction of the relative motion, will reverse the direction of the force on the electrons, reversing the polarity of the e.m.f.. **Faraday's law** states that the induced e.m.f. is proportional to the rate of change of flux linkage.

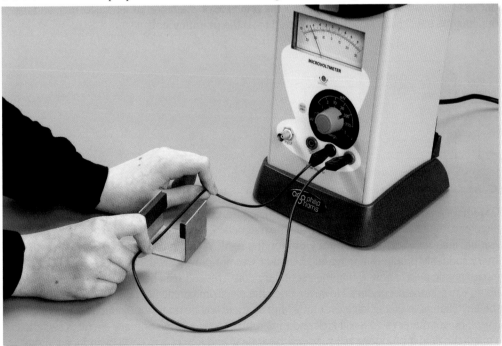

▲ **fig A** Relative movement between a wire and a magnetic field will induce an e.m.f.

PRACTICAL SKILLS

Investigating Faraday's law

You can investigate Faraday's law using a magnet and a coil of wire connected to a voltage datalogger, as shown in **fig B**, or more simply, using the set up in **fig A**.

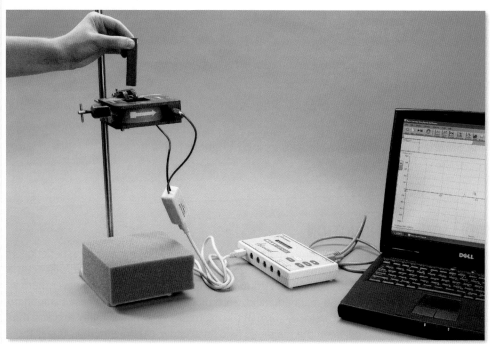

▲ **fig B** Measuring the induced e.m.f. as a magnet falls through a coil of wire.

Conservation of energy gives us **Lenz's law**, which says that 'the direction of an induced e.m.f. is such as to oppose the change creating it'.

ELECTROMAGNETIC INDUCTION USING AN ELECTROMAGNET

In the previous section, the induction of e.m.f. was a result of the relative motion between a conductor, or coil, and a permanent magnet. The magnetic field which interacts with the coil could be produced electrically by another coil. This is the principle of operation of a transformer, and the coil producing the initial magnetic field is referred to as the **primary coil**.

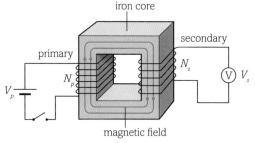

▲ **fig C** Using an electromagnet to induce e.m.f. in a coil.

In **fig C** we have a pair of coils linked together by a soft iron core. Iron is extremely good at carrying magnetism and, in the set-up shown, nearly all the magnetic field generated by the primary coil on the left would interact with the **secondary coil** on the right. When the primary switch is first closed, the primary suddenly produces a magnetic field that was not previously there. This means that a changing magnetic field is now within the secondary coil. This sudden change of flux linkage will generate an e.m.f. in the secondary. However, once the electromagnetic field is stable there will no longer be any change in flux linkage over time, and so there will be no further induced e.m.f. in the secondary. The voltmeter needle will kick and then return to zero. If the primary circuit is switched off, the magnetic field it produces will suddenly disappear, and a brief e.m.f. will be induced in the opposite direction to the switch-on voltage. The voltmeter needle will kick in the opposite direction and then return to zero again.

TRANSFORMERS

The circuit and situation described in **fig C** is not really of much practical use. It can be responsible for current surges in circuitry, which can cause damage. The same principle is more usefully applied in the transformer, in which an alternating current is supplied to the primary. As alternating current constantly varies, the electromagnetic field produced by the primary coil alternates, which creates a varying induced e.m.f. in the secondary. This e.m.f. will vary at the same rate as the current supplied to the primary. It will also change constantly depending on the varying rate of change of flux linkage. This comes from the strength of the magnetic field (which in turn depends on the number of turns on the primary coil) and the number of turns on the secondary coil. It turns out that the ratio of voltages between primary and secondary is identical to the ratio of the number of turns on these coils:

$$\frac{V_{primary}}{V_{secondary}} = \frac{N_{primary}}{N_{secondary}}$$

This gives us the essential job of a transformer, which is to change voltage. More turns on the secondary means a step-up transformer in which the output voltage is higher than the input voltage and vice versa for a step-down transformer.

EXAM HINT: EXTRA CONTENT

The transformer equation is not included in the exam specification, but it does show us factors that affect the induced e.m.f. in a set up like **fig C**.

CALCULATING INDUCED E.M.F.S

Putting Faraday's and Lenz's laws together gives us an expression for calculating an induced e.m.f.:

$$\varepsilon = -\frac{d(N\Phi)}{dt} \quad \text{or} \quad \varepsilon = -\frac{\Delta(N\Phi)}{\Delta t}$$

Faraday's law told us that the e.m.f. would be proportional to the rate of change of flux linkage, and the minus sign in the equation comes from Lenz's law, to indicate the opposing direction.

WORKED EXAMPLE 1

The iron core in **fig C** has a square cross-section with each side 4 cm. The primary produces a magnetic field with a strength of 0.5 T. This takes 60 ms to generate, and to die away, when the primary circuit is switched on or off. The secondary coil has 120 turns. What e.m.f. would be induced in the secondary at the moments of switch-on and switch-off of the primary circuit?

$$\varepsilon = -\frac{d(N\Phi)}{dt} = -\frac{\Delta(N\Phi)}{\Delta t} = -\frac{\Delta(BAN)}{\Delta t}$$

$$= -\frac{(0.5 \times 0.04^2 \times 120)}{60 \times 10^{-3}}$$

$$\varepsilon = -1.6\,V$$

At switch-off, the e.m.f. induced would be the same amount but in the opposite direction, so $\varepsilon = 1.6\,V$.

WORKED EXAMPLE 2

Zoran has a wire coil of 120 circular turns with a radius of 5 cm, held in a magnetic field with a strength of 20 mT. He removes it from within the magnetic field to a place completely outside the field in 0.02 s. What e.m.f. is induced in his coil?

Area, $A = \pi \times (0.05)^2 = 7.85 \times 10^{-3}\,m^2$

$\Phi = BA = 0.02 \times (7.85 \times 10^{-3}) = 1.6 \times 10^{-4}\,Wb$

Flux linkage = $N\Phi = 120 \times 1.6 \times 10^{-4} = 1.92 \times 10^{-2}\,Wb\text{-turns}$

This is completely *removed* in 0.02 s.

$$\varepsilon = \frac{-\Delta(N\Phi)}{\Delta t} = \frac{-(-1.92 \times 10^{-2})}{0.02}$$

$$\varepsilon = 0.96\,V$$

CHECKPOINT

1. (a) What is the flux linkage if a square coil with 10 cm sides and having 500 turns interacts with a magnetic field, $B = 0.33$ mT?

    (b) The coil is turned through 90° within the 0.33 mT magnetic field, moving from a position perpendicular to the field (full flux linkage) to a position parallel to the field (zero flux linkage). This action takes 12 ms. What is the induced e.m.f. in the coil?

2. The graph shown in **fig D** is a graph of measurements of induced e.m.f. against time as a magnet is dropped through a coil of wire.

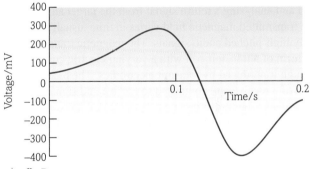

▲ **fig D**

    (a) Use the graph to describe how the induced e.m.f varies over time.

    (b) If the coil were connected in series with a light bulb, explain how the current through the bulb would change over time.

SUBJECT VOCABULARY

Faraday's law induced e.m.f. is proportional to the rate of change of flux linkage

Lenz's law the direction of an induced e.m.f. is such as to oppose the change creating it

primary coil the first coil in a transformer, through which the supply current passes

secondary coil the second coil in a transformer, through which the output current passes

METAL DETECTIVES

Using a process involving electromagnetic induction to detect metal objects was first proposed by Alexander Graham Bell, also the inventor of the telephone. Today, metal detectors are used by a variety of people in many different industries. For example, for finding underground water pipes, and by archaeologists looking for old metal objects.

▲ **fig A** Archaeologists can find buried ancient objects using electromagnetic induction.

SCIENTIFIC PAPER

METAL DETECTORS: BASICS AND THEORY

Metal detectors work on the principle of transmitting a magnetic field and analysing a return signal from the target and environment. The transmitted magnetic field varies in time, usually at rates of fairly high-pitched audio signals. The magnetic transmitter is in the form of a transmit coil with a varying electric current flowing through it produced by transmit electronics. The receiver is in the form of a receive coil connected to receive and signal processing electronics. The transmit coil and receive coil are sometimes the same coil. The coils are within a coil housing which is usually simply called 'the coil', and all the electronics are within the electronics housing attached to the coil via an electric cable and commonly called the 'control box'.

This changing transmitted magnetic field causes electric currents to flow in metal targets. These electric currents are called eddy currents, which in turn generate a weak magnetic field, but their generated magnetic field is different from the transmitted magnetic field in shape and strength. It is the altered shape of this regenerated magnetic field that metal detectors use to detect metal targets. (The different 'shape' may be in the form of a time delay.)

The regenerated magnetic field from the eddy currents causes an alternating voltage signal at the receive coil. This is amplified by the electronics because relatively deeply buried targets produce signals in the receive coil which can be millions of times weaker than the signal in the transmit coil, and so need to be amplified to a reasonable level for the electronics to be able to process.

In summary:

1 Transmit signal from the electronics causes transmit electrical current in transmit coil.

2 Electrical current in the transmit coil causes a transmitted magnetic field.

3 Transmitted magnetic field causes electrical currents to flow in metal targets (called eddy currents).

4 Eddy currents generate a magnetic field. This field is altered compared to the transmitted field.

5 Receive coil detects the magnetic field generated by eddy currents as a very small voltage.

6 Signal from receive coil is amplified by receive electronics, then processed to extract signal from the target, rather than signals from other environment magnetic sources such as Earth's magnetic field.

From a paper written by Bruce Candy, Chief Scientist, Minelab Electronics, a manufacturer of metal detectors.

https://www.minelab.com/__files/f/11043/METAL DETECTOR BASICS AND THEORY.pdf.

SCIENCE COMMUNICATION

The extract consists of information from a technical paper written for a metal detector manufacturer's customers.

1 Why is there no obvious commercial bias in the text?

2 Discuss the level of scientific detail included in the extract, and the level of language used, particularly considering the intended audience.

PHYSICS IN DETAIL

Now we will look at the physics in detail. Some of these questions will link to topics elsewhere in this book, so you may need to combine concepts from different areas of physics to work out the answers.

3 Explain how a metal detector functions, including references to Faraday's law and Lenz's law.

4 (a) Explain how the operation of the metal detector is similar to that of a transformer.

   (b) Explain how the metal detector is different from a transformer.

5 Why does the signal from the receive coil need to be amplified electronically?

6 If the target metal has a lower resistivity, it produces a stronger detection signal. Explain why a gold wedding ring might be easier to detect than a naturally occurring gold nugget of a similar size.

PHYSICS TIP

Consider how a transformer manages to transmit nearly all the power from its primary coil to its secondary coil, and then compare with the structure of this device.

ACTIVITY

Imagine you work for Minelab, and have have been asked to give a presentation about their portable metal detectors to a group of amateur treasure hunters. Prepare the part of the presentation which explains why some objects are more easily detected than others. You should cover size, shape, proximity and resistivity. Diagrams should be included in your presentation.

THINKING BIGGER TIP

Do not attempt to teach International A Level Physics to the treasure hunters. Your presentation should explain the principles without overcomplicated detail.

6C EXAM PRACTICE

[Note: In questions marked with an asterisk (*), marks will be awarded for your ability to structure your answer logically, showing how the points that you make are related or follow on from each other.]

1 Which of these is the correct SI unit for magnetic field strength?

- **A** Wb
- **B** weber-turns
- **C** T
- **D** $Wb\,m^{-3}$ [1]

(Total for Question 1 = 1 mark)

2 A beam of electrons carries 100 billion electrons 18 centimetres per second. The beam enters a magnetic field of 40 mT at right angles to the field. How much force does each electron feel?

- **A** 1.15×10^{-21} N
- **B** 1.15×10^{-19} N
- **C** 1.15×10^{-18} N
- **D** 1.15×10^{-10} N [1]

(Total for Question 2 = 1 mark)

3 When a magnet is moved into a coil of 50 turns at $0.12\,m\,s^{-1}$, an e.m.f. of 3.60 V is generated in the coil. What will be the e.m.f. generated when the magnet is removed from the coil at $0.48\,m\,s^{-1}$?

- **A** −0.90 V
- **B** 0.90 V
- **C** −14.4 V
- **D** 14.4 V [1]

(Total for Question 3 = 1 mark)

4 The diagram shows a horizontal wire which is at right angles to a magnetic field. The magnetic field is produced by a horseshoe magnet, which is on a balance adjusted to read zero when the current in the wire is zero.

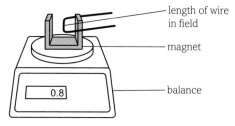

length of wire in field

magnet

0.8

balance

When the current is 4 A, the reading on the balance is 0.8 gram. The length of wire in the magnetic field is 0.05 m.
Calculate the average magnetic flux density along the length of the wire. [3]

(Total for Question 4 = 3 marks)

5 Faraday's and Lenz's laws are summarised in the list of formulae as

$$\varepsilon = -\frac{d(N\Phi)}{dt}$$

- (a) State the meaning of the term $N\Phi$. [2]
- (b) Explain the significance of the minus sign. [3]

(Total for Question 5 = 5 marks)

6 A vinyl disc is used to store music. When the disc is played, a stylus (needle) moves along in a groove in the disc. The disc rotates, and bumps in the groove cause the stylus to vibrate.

coil

vinyl disc

iron strips

magnet

stylus

The stylus is attached to a small magnet, which is near to a coil of wire. When the stylus vibrates, there is a potential difference across the terminals of the coil.

- (a) Explain the origin of this potential difference. [4]
- (b) The potential difference is then amplified and sent to a loudspeaker. Long-playing vinyl discs (LPs) have to be rotated at 33 rpm (revolutions per minute) so that the encoded bumps in the groove lead to the correct sound frequencies.
  - (i) Calculate the angular velocity of an LP. [2]
  - (ii) As the stylus moves towards the centre of the LP the encoded bumps must be fitted into a shorter length of groove.

    Explain why the encoding of bumps in the groove becomes more compressed as the stylus moves towards the centre. [3]

(Total for Question 6 = 9 marks)

7 A teacher demonstrates electromagnetic induction by dropping a bar magnet through a flat coil of wire connected to a datalogger.

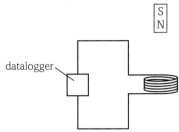

The data from the datalogger is used to produce a graph of induced e.m.f. across the coil against time.

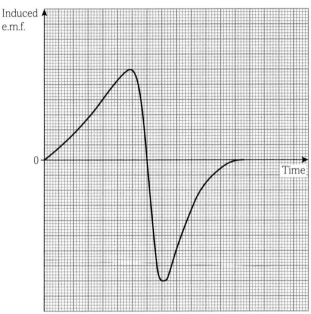

*(a) Explain the shape of the graph and the relative values on both axes. [6]

(b) The teacher then sets up another demonstration using a large U-shaped magnet and a very small coil of wire, which is again connected to a datalogger.

The north pole is vertically above the south pole and the coil is moved along the line AB, which is midway between the poles. The magnetic field due to the U-shaped magnet has been drawn. The plane of the coil is horizontal.

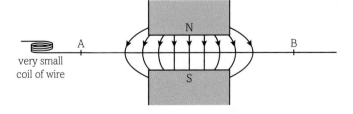

Sketch a graph to show how the e.m.f. induced across the coil varies as the coil moves from A to B at a constant speed. [4]

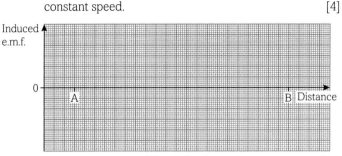

(Total for Question 7 = 10 marks)

8 Particle detectors often use magnetic fields to deflect particles so that properties of the particles can be measured.

(a) State what is meant by a magnetic field. [2]

(b) A moving charged particle of mass m, charge q and velocity v enters a uniform magnetic field of flux density B.

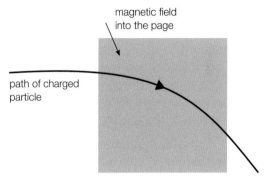

(i) State the charge on the particle. [1]

(ii) Describe and explain the shape of the particle's path in the magnetic field [2]

(iii) The radius of the path in the magnetic field is r. By considering the magnetic force acting on the particle show that the following equation is correct:

$$r = \frac{mv}{Bq}$$ [2]

(Total for Question 8 = 7 marks)

TOPIC 7 NUCLEAR AND PARTICLE PHYSICS

7A PROBING MATTER

Throughout history, people have been trying to work out what everything is made from. The ancient Greek philosopher Democritus wondered how many times a piece of cheese could be cut smaller and smaller, until we reach the point where the cheese remaining is a fundamental unit that will not split. This was the basis of the theory that everything is made from unbreakable fundamental particles.

In the late nineteenth century, the atom was thought to be the smallest possible particle, but then the English physicist J. J. Thomson demonstrated that electrons could be removed, so the atom must be made of smaller pieces. This led to some of the most important experiments of the twentieth century: Rutherford's alpha particle scattering experiments from 1909, which led to the development of the idea that an atom has a small central nucleus and outer orbiting electrons.

In this chapter, both the movement of free electrons in a beam, and the details of Rutherford's alpha particle scattering experiment will be explained. Using particle beams as probes to investigate the detailed structure of matter is one of the most important areas of physics study, and in future chapters we will see how these experiments led to the very latest developments in particle physics experimentation.

MATHS SKILLS FOR THIS CHAPTER

- Recognising and use of appropriate units (*e.g. the electronvolt*)
- Use of standard form (*e.g. calculating electron energies*)
- Use of an appropriate number of significant figures (*e.g. calculating electron velocities*)
- Changing the subject of non-linear equations (*e.g. calculating electron velocities*)
- Substituting numerical values into equations using appropriate units (*e.g. calculating de Broglie wavelengths*)

What prior knowledge do I need?

Topic 1B (Book 1: IAS)

- **Kinetic energy**

Topic 3D (Book 1: IAS)

- **Wave–particle duality, including electron diffraction**
- **The electronvolt**

Topic 4A (Book 1: IAS)

- **Electric current as the movement of electrons**
- **Voltage as a measure of the energy transferred by charges**

Topic 5A

- **Momentum conservation and elastic collisions**

Topic 6A

- **The effects of electric fields on the movements of charged particles**
- **Coulomb's law for electrostatic forces**

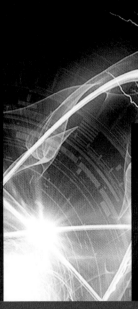

What will I study in this chapter?

- **Alpha particle deflections from nuclei**
- **Rutherford's conclusion from alpha scattering experiments**
- **The process of thermionic emission**
- **Electron beam acceleration**
- **Applications of electron diffraction and the de Broglie wavelength**

What will I study later?

Topic 7B

- **The design of particle accelerators**
- **The detection and identification of subatomic particles**

Topic 7C

- **Deeper matter probing experiments**

Topic 9A

- **The processes in which alpha particles are emitted**

LEARNING OBJECTIVES

■ Describe what is meant by *nucleon number* (*mass number*) and *proton number* (*atomic number*).
■ Explain how large angle alpha particle scattering gives evidence for a nuclear atom.
■ Describe how our understanding of atomic structure has changed over time.

ATOMIC THEORIES THROUGH HISTORY

People have always asked questions about what the materials around us consist of. There have been many philosophers and scientists, from Democritus in the fifth century BCE to Dalton in the early nineteenth century, who have suggested the idea of a tiny particle from which everything else is constructed. The basic model for one of these 'atoms' – which was published in a paper by Dalton in 1803 – is simply a hard solid sphere. When Thomson discovered that tiny negatively charged bits (electrons) could be removed from an atom, leaving behind a positively charged ion, he produced the **'plum pudding model'**. This model suggested that the main body of the atom is composed of a positively charged material (the pudding 'dough') with electrons (the 'plums') randomly scattered through it. This model was successful at explaining the evidence available at that time. In 1911 came Rutherford's model of the atom, which has a tiny, charged nucleus carrying most of the mass of the atom, surrounded at some distance by the electrons, with most of the atom as empty space. Niels Bohr later refined the nuclear model to have the electrons located around the nucleus in fixed orbits, following the theory of quantised energy. Electrons could move from one fixed orbit to another depending on the energy they gained or lost. Apart from these fixed jumps, they were stuck in an orbit. Bohr's model generated the commonly used idea of the atom as being like a miniature Solar System. This idea is not a perfect model – it is incorrect in many ways – but most models have strengths and weaknesses. It is often necessary to use models in science and we must always be careful to work within the limitations of any model we use.

LEARNING TIP

A plum pudding is a British dessert. You can imagine it like a chocolate chip muffin, where the muffin is the positive material and the chocolate chips are the electrons.

ALPHA PARTICLE SCATTERING

Between 1909 and 1911, Geiger and Marsden, students of Lord Rutherford at Manchester University, did an experiment in which they aimed alpha particles at a very thin gold foil (**fig B**).

Their expectation was that all the alpha particles would pass through, possibly with a small change in direction. The results

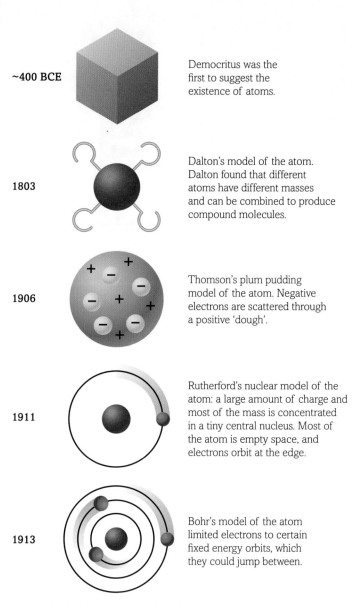

~400 BCE — Democritus was the first to suggest the existence of atoms.

1803 — Dalton's model of the atom. Dalton found that different atoms have different masses and can be combined to produce compound molecules.

1906 — Thomson's plum pudding model of the atom. Negative electrons are scattered through a positive 'dough'.

1911 — Rutherford's nuclear model of the atom: a large amount of charge and most of the mass is concentrated in a tiny central nucleus. Most of the atom is empty space, and electrons orbit at the edge.

1913 — Bohr's model of the atom limited electrons to certain fixed energy orbits, which they could jump between.

▲ **fig A** Models of the atom through history.

generally followed this pattern – the vast majority passed straight through. However, a few alpha particles had their direction changed by quite large angles. Some were even deflected back the way they had come. Rutherford commented, *'It was almost as incredible as if you had fired a 15 inch [artillery] shell at a piece of tissue paper and it came back and hit you'*. It was a shocking result based on the model of the atom at that time. They repeated the experiment hundreds of times, and Rutherford concluded that it was necessary to change the model of the atom to explain the results. He developed a model with a small nucleus in the centre of the atom which contained a large amount of charge and most of the mass of the atom (see **fig C** and **table A**).

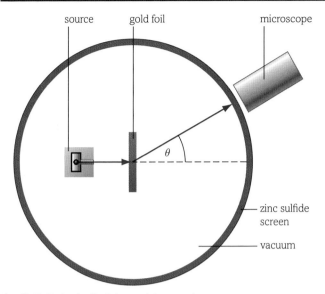

▲ **fig B** Rutherford's alpha particle scattering apparatus.

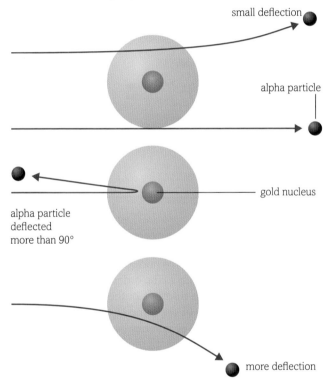

▲ **fig C** Alpha particle paths in scattering from gold atoms. Approximately 1 in 8000 was deflected through more than 90°. Momentum conservation means that this is only possible if scattering is from something more massive than the alpha particle itself.

ANGLE OF DEFLECTION/ DEGREES	EVIDENCE	CONCLUSION
0–10	Most alpha particles pass straight through with little deviation.	Most of the atom is empty space.
10–90	Some alpha particles deflected through a large angle.	A large concentration of charge in one place.
90–180	A few alpha particles are deflected back towards the source side of the foil.	Most of the mass of the atom and a large concentration of charge is in a tiny, central nucleus.

table A Alpha particle scattering observations and conclusions.

CHADWICK'S DISCOVERY OF THE NEUTRON

Rutherford determined that most of the atom's mass and all the positive charge was held in a very small nucleus in the centre, and that electrons were at the edge of the atom. The difference between the nuclear mass and the known number of protons in it caused a problem though. Nuclei were too massive considering the number of protons they contained. Rutherford suggested that additional proton–electron pairs, bound together, were the extra mass in the nucleus.

In 1930, Irene Joliot-Curie and her husband, Frederic, found that alpha particles hitting beryllium would cause it to give off an unknown radiation. Difficult to detect, this unknown, uncharged radiation could remove protons from paraffin and these were detected by a Geiger–Müller tube.

The Joliot-Curies tried to explain the unknown radiation as gamma rays, but as these rays have no mass, this was against the rules of the conservation of momentum. James Chadwick repeated the experiments (**fig D**) also using other target materials. Chadwick considered momentum transfer and conservation of kinetic energy in the collisions between the particles and he concluded that the beryllium radiation was a neutral particle which had a mass about 1% more than that of a proton. In 1932, he published a proposal for the existence of this new particle, which he called a neutron, and in 1935 he was awarded the Nobel prize for this discovery.

EXAM HINT:
EXTRA CONTENT

You will not be expected to know the details of Chadwick's experimental set-up in the exam. However, you will need to understand the idea that conservation of momentum can be applied to particles.

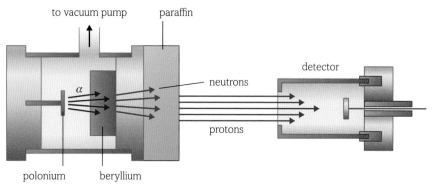

▲ **fig D** Chadwick's experiment to determine the existence of the neutron relied on the conservation of momentum to work out the mass of this neutral particle.

NUCLEAR STRUCTURE

As we have seen, the nucleus contains two types of particles: protons and neutrons. In the nucleus, these particles are all known as **nucleons**. The number of protons in a nucleus determines which element the atom will be. The periodic table is a list of the elements in the order of the number of protons in each atom's nucleus. This number is called the **proton number** or the **atomic number (Z)**. The number of neutrons can be different, and we call atoms of the same element with different numbers of neutrons, **isotopes**. For small nuclei, up to about atomic number 20 (which is calcium), the number of neutrons in the nucleus is generally equal to the number of protons.

Above atomic number 20, to be stable, more neutrons than protons are needed in the nucleus. The neutrons help to bind the nucleus together as they exert a **strong nuclear force** on other nucleons, and they act as a space between the positive charges of the protons which all repel each other. This means that as we progress through the periodic table to larger and larger nuclei, proportionately more and more neutrons are needed. By the time we reach the very biggest nuclei, there can be over 50% more neutrons than protons.

To describe any nucleus, we must say how many protons and how many neutrons there are. So the chemical symbol written below refers to the isotope of radium, which has 88 protons and 138 neutrons:

$$^{226}_{88}\text{Ra}$$

The number 226, called the **nucleon number** or the **mass number (A)**, refers to the total number of nucleons – neutrons and protons – in a nucleus of this isotope. So to find the number of neutrons, we must subtract the atomic number from the mass number: 226 – 88 = 138.

As radium must have 88 protons because that makes it radium, it is quite common not to write the 88. You might call this isotope 'radium-226' and this will be enough information to know its proton number and **neutron number**.

A QUANTUM MECHANICAL ATOM

In the 1920s, Werner Heisenberg changed the model of the atom, which had electrons in orbits like planets in a Solar System. His uncertainty principle says that we cannot know the exact position and velocity of anything at a given moment. Instead of specific orbits, his new version of the atom has regions around the nucleus in which there is a high probability of finding an electron, and the shapes of these 'probability clouds' represent what we currently refer to as the electron 'orbitals'.

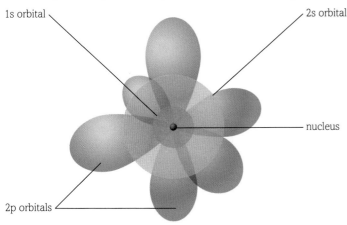

1s orbital

2s orbital

nucleus

2p orbitals

▲ **fig E** A quantum mechanical model of an atom's structure.

CHECKPOINT

1. ▶ Use a periodic table to pick five elements at random. List them using standard symbols and then explain how many protons and neutrons each one has.

2. ▶ Suggest why it has taken so long for scientists to develop the current model of the atom.

3. ▶ Give two strengths and two weaknesses of the analogy that an atom is like a mini Solar System.

4. Chadwick suggested that alpha particles colliding with beryllium produced neutrons according to the equation:

$${}^{4}_{2}\alpha + {}^{A}_{B}Be = {}^{12}_{6}C + {}^{1}_{0}n$$

What numbers are represented by A and B?

SKILLS ▶ INTERPRETATION

SKILLS ▶ PERSONAL AND SOCIAL RESPONSIBILITY

SKILLS ▶ SELF-EVALUATION

SUBJECT VOCABULARY

plum pudding model a pre-1911 model of the atom, in which the main body of the atom is made of a positively charged material (the pudding 'dough') with electrons (the 'plums') randomly scattered through it

nucleons any of the protons and neutrons comprising a nucleus

proton number the number of protons in the nucleus of an atom

atomic number (Z) an alternative name for 'proton number'

isotopes atoms of the same element with different numbers of neutrons in the nuclei

strong nuclear force the extremely short-range force between hadrons (such as protons and neutrons)*

nucleon number the total number of all neutrons and protons in a nucleus

mass number (A) an alternative name for the 'nucleon number'

neutron number the total number of neutrons within a given nucleus

* You do not need to revise this term for your exam.

LEARNING OBJECTIVES

■ Explain that electrons are released in thermionic emission.

■ Describe how electrons can be accelerated by electric and magnetic fields.

■ Explain why high energies are required to investigate the structure of nucleons.

ELECTRON BEAMS

Free conduction electrons in metals need a particular amount of energy if they are to escape from the surface of the metal. This energy can be supplied by a beam of photons, as seen in the **photoelectric effect**. The electrons can also gain enough energy through heating of the metal. The release of electrons from the surface of a metal as it is heated is known as **thermionic emission**.

LEARNING TIP

Students often confuse thermionic emission and the photoelectric effect. Whilst the two ideas have similarities, they are different and you must use the correct term with the correct situation.

If, when they escape, these electrons are in an electric field, they will be accelerated by the field, moving in the positive direction. The kinetic energy they gain will depend on the p.d., V, that they move through, according to the equation:

$$E_k = eV$$

where e is the charge on an electron.

WORKED EXAMPLE 1

How fast would an electron be moving if it was accelerated from rest through a p.d. of 2500 V?

$$E_k = eV = -1.6 \times 10^{-19} \times -2500 = 4 \times 10^{-16}\,J$$

$$E_k = \tfrac{1}{2}mv^2$$

$$v = \sqrt{\frac{2E_k}{m}} = \sqrt{\frac{2 \times 4 \times 10^{-16}}{9.11 \times 10^{-31}}}$$

$$v = 2.96 \times 10^7\,m\,s^{-1}$$

Using thermionic emission to produce electrons, and applying an electric field to accelerate them, we can generate a beam of fast-moving electrons, known as a **cathode ray**. If this beam of electrons passes through a further electric field or magnetic field, then the force produced on the beam of electrons will cause it to deflect. If a fast-moving electron hits a screen that is painted with a particular chemical, the screen will fluoresce – it will emit light.

These are the principles by which cathode ray oscilloscopes (CROs) operate. The electron beam in a CRO is moved left and right, and up and down, by passing the beam through horizontal and vertical electric fields. These are generated by electric plates so the strength and direction can be altered. The point on the screen which is emitting light can be changed quickly and easily.

AN ELECTRON PROBE

Electron beams fired at a crystal will produce scattering patterns that can tell us about the structure of the crystal (**fig A**). In 1927, Davisson and Germer showed that an electron beam can produce a diffraction pattern. This was different to the patterns found by Geiger and Marsden in Rutherford's alpha particle scattering experiments. Davisson and Germer provided the experimental evidence to prove a theory that had been suggested three years earlier by the French physicist, Louis de Broglie. Light could be shown to behave as a wave sometimes, and at other times as a particle. He hypothesised that this might also be the case for things which were traditionally thought to be particles. De Broglie had proposed that the wavelength, λ, of a particle could be calculated from its momentum (p) using the expression:

$$\lambda = \frac{h}{p}$$

where h is the Planck constant. So:

$$\lambda = \frac{h}{mv}$$

The Davisson–Germer experiment proved that the diffraction pattern obtained when a cathode ray hit a crystal could only be produced if the electrons in the beam had a wavelength that was the de Broglie wavelength. Because of this experimental confirmation, Louis de Broglie was awarded the 1929 Nobel Prize for Physics.

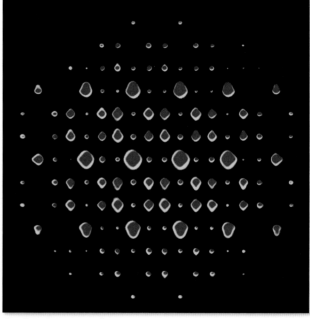

▲ **fig A** Electron diffraction patterns can explore the molecular structure of crystals.

WORKED EXAMPLE 2

What is the wavelength of an electron in a beam which has been accelerated through 2000 V?

$$E_k = eV = -1.6 \times 10^{-19} \times -2000 = 3.2 \times 10^{-16} \, J$$

$$E_k = \tfrac{1}{2} mv^2 = 3.2 \times 10^{-16} \, J$$

$$\therefore \quad v = \sqrt{\frac{2E_k}{m}} = \sqrt{\frac{2 \times 3.2 \times 10^{-16}}{9.11 \times 10^{-31}}} = 2.65 \times 10^7 \, m\,s^{-1}$$

$$\lambda = \frac{h}{mv} = \frac{6.63 \times 10^{-34}}{9.11 \times 10^{-31} \times 2.65 \times 10^7}$$

$$\lambda = 2.75 \times 10^{-11} \, m$$

EXAM HINT

Many standard values, like the charge and mass of an electron and the Planck constant, are given in the data sheet in the exam.

PRACTICAL SKILLS

Investigating electron diffraction

You may have the equipment to observe electron diffraction. By measuring the radius of the circular pattern for each accelerating voltage, you can perform a calculation to confirm de Broglie's hypothesis.

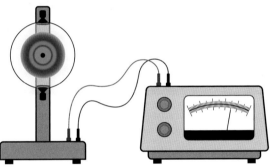

▲ **fig B** Measuring electron diffraction caused by a very thin piece of carbon.

⚠️ Safety Note: The power supplies can give a severe electric shock and should be used with shrouded connectors. Do not remove or attach connectors with the power switched on.

The idea of electrons acting as waves has enabled scientists to study the structure of crystals, the same way they do in X-ray crystallography. When waves pass through a gap which is about the same size as their wavelength, they are diffracted. In other words, they spread out. The degree of diffraction spreading depends on the ratio of the size of the gap to the wavelength of the wave. If a beam of electrons is aimed at a crystal, the gaps between atoms in the crystal can act as a diffraction grating and the electron waves produce a diffraction pattern on a screen. Measuring the pattern allows the spacings between the atoms to be calculated.

Electron diffraction and alpha particle scattering both show the idea that we can study the structure of matter by probing it with beams of high energy particles. The more detail (or smaller scale) the structure to be investigated has, the higher energy the beam of particles needs to be. This means that very high energies are needed to investigate the structure of nucleons, as they are very, very small. Accelerating larger and larger particles to higher and higher energies has been the aim of particle physicists ever since Thomson discovered the electron in 1897.

CHECKPOINT

SKILLS   SELF-DIRECTION, ADAPTIVE LEARNING, INTERPRETATION, COMMUNICATION, ADAPTABILITY

1. Calculate the de Broglie wavelengths for the following:
   (a) an electron travelling at 2% of the speed of light
   (b) an electron which has been accelerated through 1200 V
   (c) a proton with a momentum of 5×10^{-21} kg m s^{-1}
   (d) you running at 5 m s^{-1}.

2. Why would de Broglie not have been awarded the Nobel Prize before the Davisson–Germer experiment?

3. Calculate the speed at which an electron must travel if it is to be used to probe the structure of the nucleus. (It would need a de Broglie wavelength of about the size of the nucleus: $\lambda = 5 \times 10^{-15}$ m.) Comment on your answer.

4. ▶ Carry out some research to find out how the direction of the electron beam in a cathode ray oscilloscope (CRO) can be changed in order to make any point on the screen light up. Show this in a diagram. Explain how such a CRO could be set up in a hospital to display the electrical impulses of a patient's heart.

SUBJECT VOCABULARY

photoelectric effect when electrons are released from a metal surface because the metal is hit by electromagnetic radiation

thermionic emission the release of electrons from a metal surface caused by heating of the metal

cathode ray a beam of electrons

XENON-XENON

The Large Hadron Collider at CERN has undertaken some very high energy experiments, including the acceleration of whole atoms, as compared with individual particles, like protons, when it first started.

ONLINE NEWSPAPER ARTICLE

A DAY OF XENON COLLISIONS AT CERN

The image to the right shows what happened in the CMS particle detector when xenon nuclei were circulated in the LHC and brought into head-on collision. The yellow is made up of tracks of electrically-charged particles, produced in such numbers that the whole of the centre of the picture is a yellow blur, with individual tracks only visible near the edges. The blue and green blocks indicate energy deposited by both charged and neutral particles in the CMS calorimeter.

Collisions between protons look significantly less busy than this, with fewer particles produced. But both xenon and lead nuclei are packed with protons and neutrons, and though lead has more of them, I don't think anyone could tell the difference between a xenon-xenon collision and a lead-lead one by eye.

There are however expected to be differences in detail, on average, in the shape and properties of the exotic ball of material produced in the heart of the collisions.

Before it flies apart, this material is a plasma of quarks and gluons, the basic constituents of all nuclear material. Measuring the differences between lead collisions and xenon collisions may teach us more about this strange stuff. If nothing else, it should allow us to make some "control" measurements, a good way of reducing systematic uncertainties. And measuring something new could always throw up a surprise. Time will tell, as the recorded data are carefully analysed.

In a coincidental aside, a xenon nucleus contains 54 protons, and about 77 neutrons on average, which makes its total mass quite close to the mass of the Higgs boson. This coincidence has caused confusion in some circles.

There should be no confusion really. According to the Standard Model, the Higgs boson is infinitely small. We can't measure "infinitely small", of course, but from the way it behaves, we can set some kind of upper limit on the physical size of a Higgs. This tells us that if the xenon nucleus were the size of a beachball, the Higgs boson would be smaller than the finest grain of sand.

This would go some way toward explaining why the Higgs boson was not discovered until 2012, whereas xenon was discovered in 1898. Another reason would be that the Higgs boson rapidly and spontaneously decays into other particles, whilst xenon is stable.

Unless, of course, you treat xenon like we did in the LHC.

From *https://www.theguardian.com/science/life-and-physics/2017/oct/15/a-day-of-xenon-collisions-at-cern*

SCIENCE COMMUNICATION

The article is from a UK-based newspaper, *The Guardian*.

1 (a) Describe the level of science presented in this article, and explain how scientific concepts are presented for the intended audience. Your answer should include commentary about the style of the text.

   (b) Comment on the relevance of the picture accompanying the article.

2 The newspaper tries hard to have an international presence and make its website popular in many countries. Discuss how the level of English language used in the article would help this, or be a barrier to readers with English as an additional language.

PHYSICS IN DETAIL

Now we will look at the physics in detail. Some of these questions will link to topics elsewhere in this book, so you may need to combine concepts from different areas of physics to work out the answers.

3 Write a paragraph to explain the structure of the xenon nuclei being used, and how this compares with lead nuclei previously used.

4 How would xenon nuclei and lead nuclei be affected if they moved across a magnetic field? Explain the differences in effect on these two particles travelling through the same magnetic field.

5 Explain the physics behind the following phrases from the extract:

   (a) 'Collisions between protons look significantly less busy than this'

   (b) '"control" measurements, a good way of reducing systematic uncertainties'

6 Give an order of magnitude approximation of the ratio of the diameters of a xenon nucleus and a Higgs boson.

ACTIVITY

Write a script for a *Guardian* podcast of this news story. You should include the same main basic information as the website article, and develop the depth of the scientific reporting for an adult audience who are likely to be well-educated but not experts in particle physics. You should also include some information about the results achieved so far. This task may require some further research.

[Note: In questions marked with an asterisk (*), marks will be awarded for your ability to structure your answer logically, showing how the points that you make are related or follow on from each other.]

1 Electrons are accelerated from rest to a kinetic energy of 2.50 keV.

(a) Calculate the potential difference that the electrons were accelerated through.

   A 2.50 V

   B 2500 V

   C 1.56×10^{19} V

   D 1.56×10^{22} V   [1]

(b) Calculate the energy of these electrons in joules.

   A 4.00×10^{-19} J

   B 4.00×10^{-16} J

   C 1.56×10^{19} J

   D 1.56×10^{22} J   [1]

(Total for Question 1 = 2 marks)

2 Calculate the speed of electrons which have a kinetic energy of 3000 eV.

   A 7.39×10^{-14} m s^{-1}

   B 2.30×10^{7} m s^{-1}

   C 3.25×10^{7} m s^{-1}

   D 1.05×10^{15} m s^{-1} [1]

(Total for Question 2 = 1 mark)

3 Heating a metal wire can cause electrons to escape from the metal. This process is called:

   A electron excitation

   B fluorescence

   C the photoelectric effect

   D thermionic emission.   [1]

(Total for Question 3 = 1 mark)

*4 Rutherford designed an experiment to see what happened when alpha particles were directed at a piece of gold foil. Summarise the observations and state the conclusions that Rutherford reached about the structure of gold atoms.   [5]

(Total for Question 4 = 5 marks)

5 The de Broglie wave equation can be written $\lambda = \sqrt{\dfrac{h^2}{2mE_k}}$,

where m is the mass of a particle and E_k is its kinetic energy.

(a) Derive this equation. Use the list of equations provided on the formulae sheet (Appendix 7 of the specification). [2]

(b) An electron is accelerated through a potential difference of 2500 V.

Using the equation $\lambda = \sqrt{\dfrac{h^2}{2mE_k}}$, calculate the de Broglie wavelength of this electron. [3]

(Total for Question 5 = 5 marks)

6 In an experiment to investigate the structure of the atom, α-particles are fired at a thin metal foil, which causes the α-particles to scatter.

(a) (i) State the direction in which the number of α-particles detected will be a maximum. [1]

   (ii) State what this suggests about the structure of the atoms in the metal foil. [1]

(b) Some α-particles are scattered through 180°. State what this suggests about the structure of the atoms in the metal foil. [2]

(c) The diagram shows the path of an α-particle passing near to a single nucleus in the metal foil.

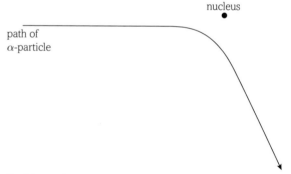

(i) Name the force that causes the deflection of the α-particle. [1]

(ii) On a copy of the diagram, draw an arrow to show the direction of the force acting on the α-particle at the point where the force is a maximum. Label the force F. [2]

(iii) The foil is replaced by a metal which has greater proton number.

Draw the path of an α-particle that has the same initial starting point and velocity as the one which is drawn in the diagram. [2]

(Total for Question 6 = 9 marks)

7 Draw a diagram and explain: (i) the structure of the atom that was proposed after the results from Rutherford's alpha particle scattering experiment were published, and (ii) the evidence for this model of the atom.   [6]

(Total for Question 7 = 6 marks)

8 (a) 'Fast moving electrons with a de Broglie wavelength of 0.11 nm are sometimes used to investigate the atomic structure of solids, such as rubber.'

Describe and explain what you can conclude from this statement about the wavelength of the electrons used?   [1]

(b) (i) Write the equation to calculate the de Broglie wavelength for a particle.   [1]

(ii) Calculate the momentum of electrons with this de Broglie wavelength.   [2]

(iii) Calculate the velocity of electrons with this momentum or wavelength.   [1]

(iv) Calculate the potential difference through which the electrons would have to be accelerated to give them this momentum or wavelength.   [2]

(c) 'A stream of neutrons could be used instead of the electrons.'

Explain how the speed of neutrons with a de Broglie wavelength of 0.11 nm would compare with that of the electrons used in part (b).   [3]

(Total marks for question 8 = 10 marks)

9 The electron in a hydrogen atom can be described by a stationary wave which is confined within the atom. This means that the de Broglie wavelength associated with it must be similar to the size of the atom, which is of the order of 10^{-10} m.

(a) (i) Calculate the speed of an electron whose de Broglie wavelength is 1.00×10^{-10} m.   [3]

(ii) Calculate the kinetic energy of this electron in electronvolts.   [3]

(b) When ß radiation was first discovered, it was suggested that there were electrons in the atomic nucleus, but it was soon realised that this was impossible because the energy of such an electron would be too great.

Explain why an electron confined within a nucleus would have a much greater energy than the energy calculated in part (a)(ii).   [2]

(Total for Question 9 = 8 marks)

TOPIC 7 NUCLEAR AND PARTICLE PHYSICS

PARTICLE ACCELERATORS AND DETECTORS

It is remarkable that scientists have managed to find out the nature of particles that measure just one femtometre across, that is the size of an atomic nucleus. To help you understand how small that is, imagine 1000 solar systems next to each other. Now compare your size to that of the thousand solar systems. That difference is the same as the size of the proton compared to you.

Developing both the theories and the technology to move subatomic particles so that we can do experiments with them is one of the great triumphs of modern science. The Large Hadron Collider, which is our greatest success story in such tiny-scale experiments, is actually the largest machine ever built!

In this chapter, we will look at how we can make tiny particles move in such a way that we can make them collide with each other, so they interact and produce other particles. To discover what particles these collisions generate, we will also learn how they can be detected and then identified.

MATHS SKILLS FOR THIS CHAPTER

- **Recognising and use of appropriate units** (*e.g. the electronvolt*)

- **Making order of magnitude calculations** (*e.g. comparing cyclotron energies*)

- **Estimating results** (*e.g. comparing particle speeds in different accelerators*)

- **Changing the subject of non-linear equations** (*e.g. calculating relativistic synchroton frequency*)

- **Substituting numerical values into equations using appropriate units** (*e.g. calculating particle path radii*)

What prior knowledge do I need?
Topic 4A (Book 1: IAS)
- Voltage as a measure of the energy transferred by charges
Topic 5A
- The conservation of momentum, charge and energy
Topic 5B
- Circular motion and centripetal force
Topic 6A
- The effects of electric fields on the movement of charged particles
- Coulomb's law for electrostatic forces
Topic 6C
- The effects of magnetic fields on the movement of charged particles

What will I study in this chapter?
- The structure and functioning of linear particle accelerators
- The structure and functioning of circular particle accelerators
- The detection and identification of subatomic particles
- How to interpret particle tracks in particle detectors
- Deeper matter probing experiments, including the Large Hadron Collider

What will I study later?
Topic 7C
- The various fundamental particles that have been discovered
- Interactions between fundamental particles
- The four fundamental forces of nature
Topic 9A
- Nuclear fission and nuclear fusion
Topic 11B
- The importance of particles in stars and between stars

STOP
NOT SOLID
DO NOT STEP!

LEARNING OBJECTIVES

■ Describe the use of electric and magnetic fields in particle accelerators.

■ Derive and use the equation $r = \dfrac{p}{BQ}$ for a charged particle in a magnetic field.

■ Explain why high energies are required to investigate the structure of the nucleus.

PARTICLE ACCELERATORS

To investigate the internal structure of particles like nucleons – protons and neutrons – scientists collide them with other particles at very high speeds (very high energies). It is necessary to use high energy particles because at lower energies the particles just bounce off each other, keeping their internal structure secret. If we can collide particles together hard enough, they will disintegrate, and show their structure. In most cases, additional particles are created from the energy of the collision. Sometimes these extra particles uncover or confirm other new physics.

The challenge for scientists has been to accelerate particles to high enough speeds. Charged particles can be accelerated in straight lines using electric fields, and their direction changed along a curved path by a magnetic field.

LINEAR ACCELERATORS

One of the simplest ways to produce energy high enough for these particle collisions is to accelerate a beam of charged particles along a straight path. However, the maximum achievable potential difference limits this. To overcome this problem, the particles are accelerated in stages (**fig A**). They are repeatedly accelerated through the maximum p.d., making the particle energies very high. Using this principle, the 3.2 km Stanford Linear Accelerator in the USA can accelerate electrons to an energy of 50 GeV, meaning that they have effectively passed through a potential difference of 50 billion volts. 'GeV' means gigaelectronvolts, or '$\times 10^9$ eV'.

DID YOU KNOW?

LINAC FACTS

The Stanford Linear Accelerator in California is still the world's longest linear accelerator or LINAC, despite being 50 years old. However, plans for the International Linear Collider are far advanced. Most likely to be built in Japan, this would consist of two linear colliders, each over 15 km long, designed to collide electrons and positrons, initially with energies of 250 GeV, but with the design possibility of raising this to 1 TeV.

If the particles to be accelerated by the **linear accelerator** in **fig A** are electrons, they are generated by an electrostatic machine (like a Van de Graaff generator) and then put into the machine. Once inside the cylinder, the electrons move in a straight line, as the electrode is equally attracting in all directions. The alternating voltage supply is made to change as the electrons reach the middle of tube A, so it becomes negative. This repels the electrons out of the end of tube A and on towards tube B, which now has a positive potential. They accelerate towards it, and the whole process repeats as they pass through tube B and are then accelerated on towards tube C. This carries on until the electrons reach the end of the line, at which point they emerge to collide with a target.

In order to keep accelerating particles that are moving faster and faster, the acceleration tubes must be made longer and longer as the particles travel through each successive one at a higher speed. The time between potential difference flips is fixed as the alternating voltage has a uniform frequency of a few gigahertz (often referred to as radio frequency, RF). Using this type of accelerator is limited by how long you can afford to build it. The whole structure must be in a vacuum so that the particles do not collide with air atoms, and it must be perfectly straight.

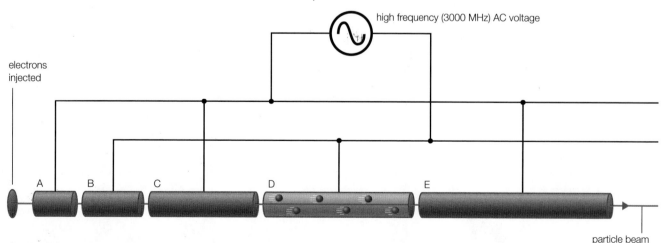

high frequency (3000 MHz) AC voltage

electrons injected

A B C D E

particle beam

▲ **fig A** The structure of a linear accelerator..

WAS EINSTEIN RIGHT?

One of Einstein's claims in his theory of special relativity is that nothing can accelerate beyond the speed of light. This means that particles in accelerators must have a problem when they are already travelling close to the speed of light and then pass through a p.d. which should accelerate them beyond it. In 1964, William Bertozzi demonstrated that at very high speeds, particles deviate from the equation $\frac{1}{2}mv^2 = qV$ and never accelerate beyond the speed of light.

By measuring the actual speed of electrons accelerated from a Van de Graaff generator and then determining their actual kinetic energy by colliding them with a target and measuring the heat generated, Bertozzi was able to show that the energy added by the accelerating p.d. started to become more than the amount expected from $\frac{1}{2}mv^2$.

This demonstration showed that whilst the kinetic energy and momentum of particles can continue to increase without limit, their speed does not. This can only happen if the mass of a particle seemingly increases with its speed. This apparent mass increase becomes significant at speeds approaching light speed – known as 'relativistic speeds'.

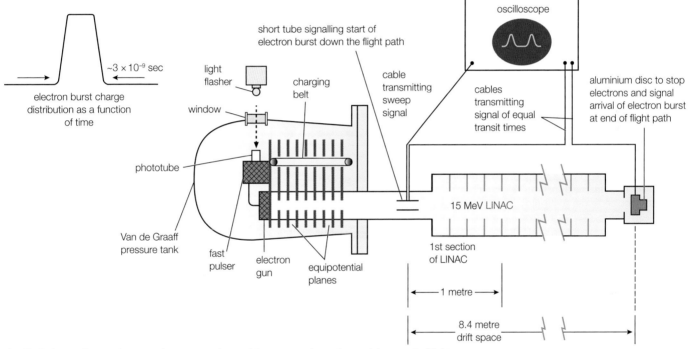

▲ **fig B** Bertozzi's experiment to demonstrate that nothing can accelerate beyond the speed of light.

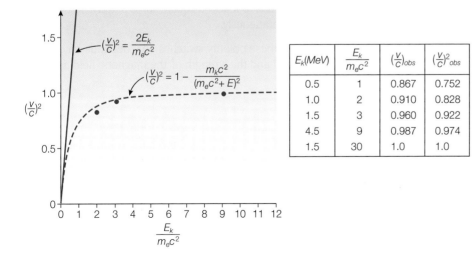

$E_k(MeV)$	$\dfrac{E_k}{m_e c^2}$	$\left(\dfrac{v}{c}\right)_{obs}$	$\left(\dfrac{v}{c}\right)^2_{obs}$
0.5	1	0.867	0.752
1.0	2	0.910	0.828
1.5	3	0.960	0.922
4.5	9	0.987	0.974
1.5	30	1.0	1.0

◀ **fig C** The solid curve represents the prediction for $\left(\frac{v}{c}\right)^2$ according to Newtonian mechanics, $\left(\frac{v}{c}\right)^2 = \frac{2E_k}{m_e c^2}$. The dashed curve represents the prediction of special relativity, $\left(\frac{v}{c}\right)^2 = 1 - \frac{m_e c^2}{(m_e c^2 + E_k)^2}$.

m_e is the rest mass of an electron and c is the speed of light in a vacuum, $3 \times 10^8\,\text{m s}^{-1}$. The solid circles are the data of Bertozzi's experiment. The table presents the observed values of $\left(\frac{v}{c}\right)$.

ACCELERATING PARTICLES IN CIRCLES

Scientists found it difficult to make linear accelerators which were longer and longer so they started to coil the accelerators up in a circle. This means that the particles could be accelerated in an electric field repeatedly in a smaller space. To do this, we use the fact that charged particles moving across a magnetic field will experience a centripetal force, and so will move in a circular path. We can work out the radius of this circular path, and use it to construct a circular accelerator of the right dimensions.

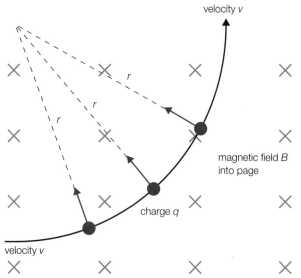

▲ **fig D** The circular trajectory of a charged particle moving across a magnetic field. Note that field lines can be shown coming out of a page by drawing a dot, and those going into a page (like the magnetic field here) can be shown by drawing a cross.

We have previously seen the equation for the force on a charged particle moving across a magnetic field (**Section 6C.3**):

$$F = Bqv$$

This force acts at right angles to the velocity, v, meaning that the particle will follow a circular path. Remember that the equation for the centripetal force on anything moving in a circle (**Section 5B.2**) is:

$$F = \frac{mv^2}{r}$$

We can equate these two expressions:

$$Bqv = \frac{mv^2}{r}$$

Dividing out the velocity from each side, and rearranging to find an expression for the radius of the circle, gives:

$$r = \frac{mv}{Bq}$$

or: $$r = \frac{p}{Bq}$$

This means that for a given magnetic field, the radius of the path of a charged particle is proportional to its momentum. At slow speeds, the radius is proportional to velocity (or the square root of the kinetic energy). However, these experiments generally send particles at speeds approaching the speed of light, so relativistic effects need to be accounted for. In particular, at these very high speeds the particle's mass appears to increase, which would also change its momentum. The overall result is that a particle increasing in speed would travel along an outwardly spiralling path.

THE CYCLOTRON

In 1930, Ernest Lawrence developed the first **cyclotron**. This was a circular accelerator which could give protons about 1 MeV of energy.

In a cyclotron, there are two D-shaped electrodes (or dees), and the particles are accelerated in the electric field in the gap between them. Within the dees, the particle will travel along a semicircular path because of the magnetic field, before being accelerated across the gap again; then another semicircle, another acceleration across the gap, and so on. As each acceleration increases the momentum of the particle, the radius of its path within the dee increases, and so it steadily spirals outwards until it emerges from an exit hole and hits the target placed in a bombardment chamber in its path.

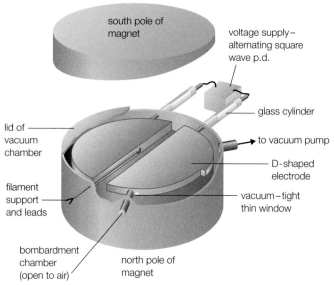

▲ **fig E** The structure of a cyclotron.

CYCLOTRON FREQUENCY

In order to maintain the accelerations at exactly the correct time, the p.d. needs to switch direction exactly when the particle exits from one dee to move across the gap between them. This means the voltage supply has to follow a square wave pattern so that it flips polarity instantaneously.

The frequency of these polarity switches only depends on the particle being used and the strength of the magnetic field applied:

$$f = \frac{1}{T}$$

$$T = \frac{2\pi r}{v}$$

During one complete period of the alternating voltage the particle will pass through both dees, thus completing a full circle at that radius.

However:

$$r = \frac{mv}{Bq}$$

$$\therefore \quad T = \frac{2\pi mv}{Bqv}$$

$$T = \frac{2\pi m}{Bq}$$

$$\therefore \quad f = \frac{Bq}{2\pi m}$$

So the frequency needed is independent of the radius. This means that a constant frequency can be used and the particle will complete each semicircle through a dee in the same time. That is, until the speed becomes so fast that mass changes through relativistic effects!

THE SYNCHROTRON

Once the relativistic effects of mass increase are included in the equation for cyclotron frequency, it becomes:

$$f = \frac{Bq}{2\pi m_0} \times \sqrt{1 - \frac{v^2}{c^2}}$$

Note that m_0 is the particle's rest mass. This shows that the frequency of the applied accelerating potential difference depends on the velocity of the particles. So a cyclotron needs very complex circuitry which produces accurately timed polarity switches to enable it to generate high energy particle beams. The necessary electronics were first developed in 1945 and the 'synchrocyclotron' can accelerate particles to as much as 700 MeV.

With the development of varying frequency for the accelerating p.d., the next step was to vary the strength of the magnetic field using an electromagnet. This means that the radius of the particle beam's path could be kept constant by increasing the magnetic field strength in line with the increasing momentum. A single ring accelerator like this is called a **synchrotron**.

Alternate accelerating tubes and bending magnets can generate very high energy particle beams. For example, the Tevatron at Fermilab near Chicago, in the USA, could give particles 1 TeV of energy (Tera = 10^{12}). Alternatively, these rings can be used to store charged particles by making them circulate endlessly at a constant energy. This is a particularly important use when trying to store anti-matter, which will annihilate if it comes into contact with normal matter (see **Section 7C.1**).

NINOVIUM: SCIENTISTS SOMETIMES CHEAT TOO

In 1999, the Lawrence Berkeley Laboratory in California published a paper claiming that they had produced an isotope of an element with atomic number 118. The paper claimed that they had succeeded in observing a chain of alpha decays following their collision of 449 MeV krypton-86 ions into a target of lead-208. The radioactive decay sequence was indicative of having started from element 118 – initially named 'ninovium' after the Bulgarian team leader Victor Ninov. No other lab in the world was able to reproduce the results, and after a two-year inquiry, it was found that Ninov had falsified the results. The Berkeley Lab retracted the paper, and Victor Ninov, who has continually maintained his innocence, was sacked.

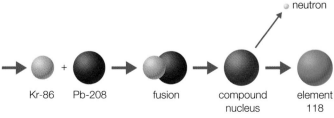

▲ **fig F** The fusion reaction which was claimed to produce element 118.

In 2006, an accelerator facility in Dubna, Russia, working jointly with an American group from the Lawrence Livermore National Laboratory, claimed to have fused calcium-48 and californium-249 to produce Oganesson, element 118, with a mass number of 294.

The team used a particle accelerator to hit Cf-249 with high energy ions of Ca-48. With the previous problems surrounding this heavy element 118, the research team were very careful about the accuracy of their data, but it took ten years before the results were accepted as valid by the International Unions of Pure and Applied Chemistry, and of Pure and Applied Physics (IUPAC and IUPAP). Mark Stoyer, a nuclear chemist at Lawrence Livermore National Laboratory, who was part of the research team, said:

'*We saw something interesting, and the way we've interpreted it, it's element 118. But you need someone else to duplicate it. A fundamental tenet of science is reproducibility.*'

CHECKPOINT

SKILLS ▷ PROBLEM SOLVING

1. Why do the electrodes in a linear accelerator get progressively longer?

2. Lawrence's second cyclotron was 25 cm in diameter and accelerated protons to 1 MeV.
   (a) Calculate the speed of these protons.
   (b) Calculate the momentum of these protons
   (c) Calculate the magnetic field strength Lawrence used in this cyclotron, assuming the protons move in a circle around the very edge of the machine.
   (d) Calculate the frequency of the voltage that Lawrence had to apply to achieve this proton acceleration.

3. ▷ The Large Hadron Collider (LHC) is a giant circular accelerator near Geneva in Switzerland. The LHC website has some facts about it:
   • the circular ring has a circumference of 27 km
   • each proton goes around over 11 000 times a second
   • each proton has 7×10^{12} eV of kinetic energy.
   (a) Calculate the speed of these protons.
   (b) Calculate $\frac{1}{2}mv^2$ for these protons using $mp = 1.67 \times 10^{-27}$ kg.
   (c) How does your answer to (b) compare with the 7 TeV of kinetic energy the protons are given through p.d. acceleration?
   (d) Explain the difference.

SUBJECT VOCABULARY

linear accelerator a machine which accelerates charged particles along a straight line

cyclotron a circular machine that accelerates charged particles, usually following a spiral path

synchrotron a machine that accelerates charged particles around a fixed circular path

EXAM HINT: EXTRA CONTENT

For the exam, you are only required to know about the linear accelerator and the cyclotron. You will not be asked questions about the synchrotron in the exam.

LEARNING OBJECTIVES

■ Describe the roles of electric and magnetic fields in particle detectors.
■ Apply conservation of charge, energy and momentum to interactions between particles and interpret particle tracks.

PRINCIPLES OF DETECTION

Both school laboratory electron diffraction and Geiger and Marsden's alpha scattering experiment rely on observing light given off from a fluorescent screen when it is hit by a particle. In their alpha particle scattering experiment, Geiger and Marsden sat for two years in a dark room, for eight hours at a time, counting the flashes of light they saw through a microscope. Geiger found this so frustrating that he jointly invented the Geiger–Müller (GM) tube to detect particles, which could then be counted electronically.

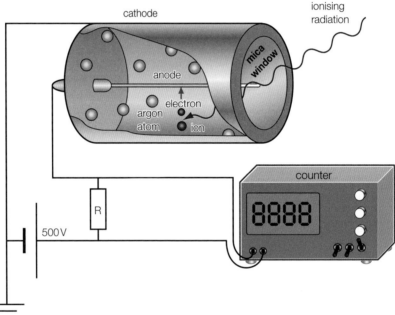

▲ **fig A** The design of a Geiger–Müller detector.

EXAM HINT: EXTRA CONTENT

For the exam, you are only required to know the general principles of the ionisation and electric fields used in such detectors. You do not have to know the details of the design for a GM tube.

The GM tube works on the principle that is common to most particle detectors: ionisation. As the particle to be detected passes through the tube, it ionises atoms of a gas (typically argon) which fills the tube. The ions and electrons produced are accelerated by an electric field between electrodes in the tube and then discharged when they reach the electrodes. This produces a pulse of electricity, which is counted by a counter connected to the tube. Many different types of detector have been invented by particle physicists, but the majority detect ionisation caused by the particles to be detected.

ANALYSING DETECTIONS

Particle-counting detectors are useful, but they cannot distinguish between different types of particles. This is becoming very important in detectors. Modern particle physics experiments are carried out using such high energies that they can produce hundreds of different types of particles. Unless the detectors used in these experiments can identify properties of the particles, such as their energy, charge or mass, then the experimental results will be useless. In **Section 6C.3** we saw how the mass spectrometer deflects an ion using a magnetic field. The ion can then be identified by the amount of deflection, since this is dependent on the mass and charge of the ion, and the strength and direction of the magnetic field used.

▲ **fig B** A hydrogen bubble chamber.

Particle physics experiments took a major leap forward with the invention of the **bubble chamber** detector (**fig B**). Professor Siegbahn of the Swedish Academy of Sciences commented when presenting the 1960 Nobel Prize in Physics:

> *"Dr. Glaser, your invention of the 'bubble chamber' has opened up a new world for nuclear science. We can now observe with our own eyes all the strange processes which occur when high-energy beams from [particle accelerators] are directed into your chamber."*

Professor Siegbahn went on to explain that the bubble chamber acts like a combination of jet-plane vapour trails and the bubbles that suddenly appear when you open a bottle of fizzy drink. Superheated liquid hydrogen changes into bubbles of gas at any place where ions are generated within it. These bubbles can be observed within the liquid and the trails of bubbles show the paths of moving particles.

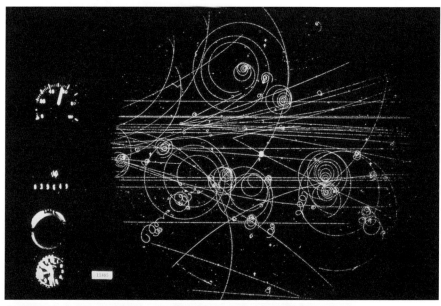

▲ **fig C** An image of particle tracks detected by a hydrogen bubble chamber.

The particle tracks in **fig C** show how particles can be tracked as they progress through a bubble chamber. Those affected by the magnetic field across it follow curved paths. The radius of curvature of the tracks will tell us the mass and charge of the particles in the same way that it does in the mass spectrometer. In addition, the picture allows us to analyse interactions as they happened, as the tracks sometimes end abruptly or have sharp changes in direction when they collide. In some cases, particle tracks appear to start from nothing. These show instances where particles have been created from uncharged particles that do not show.

WORKED EXAMPLE 1

Look at track A in **fig D**. If the magnetic field in the bubble chamber went at right angles to this picture and into the page, would the particle shown by track A be an electron or a positron (positively charged anti-electron)?

▲ **fig D** Fleming's left hand rule acting on the particle of track A.

Using Fleming's left hand rule, with the index finger pointing into the page (magnetic field), the centripetal force shown by the thumb leaves the middle finger, indicating a current flowing, along the spiral inwards from the larger outside loops. The particle would be slowing down as it loses kinetic energy by ionising other particles in its path. The radius of curvature for these tracks is proportional to the velocity of the particle $\left(r = \dfrac{mv}{Bq}\right)$ so deceleration would cause the particle to spiral inwards. This means the particle must be travelling in the same direction along the track as the flow of conventional current, as defined in Fleming's left hand rule, so the particle must be positive – a positron.

CHECKPOINT

SKILLS PRODUCTIVITY, REASONING/ARGUMENTATION

1. Explain why ionisation is important in the operation of the Geiger–Müller tube.

2. In **fig E**, the red and yellow tracks show the creation of an electron and positron. Identify which track is the electron. The magnetic field goes into the page.

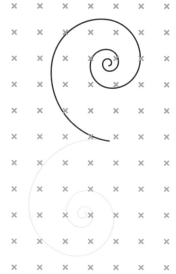

▲ **fig E** The creation of an electron–positron pair.

3. ▶ Fig F was taken by Carl Anderson and published in *Physical Review* in March 1933 as evidence for a positively charged particle with the mass of an electron. He had discovered the positron.

▲ **fig F** Anderson's positron track is the thin curving grey line.
The dark grey bar across the middle of the picture is a lead absorber that removes about two thirds of the particle's energy. Explain how you can tell that the particle is moving upwards in the picture.

SUBJECT VOCABULARY

bubble chamber a particle detection system in which the particles cause bubbles to be created in a superheated liquid, typically hydrogen

CERN'S BIG SYNCHROTRON

▲ **fig A** Inside the Large Hadron Collider.

At the Organisation Européenne pour la Recherche Nucléaire (known by its older acronym CERN), or European Organisation for Nuclear Research, they have constructed the biggest machine in the history of planet Earth. This is the Large Hadron Collider (LHC) – a giant synchrotron over 8 km in diameter, built 100 metres underground on the border between Switzerland and France. At a total building cost of just under 4 billion euros, with staff from 111 nations involved in its building and operation, it is the largest experiment ever undertaken. When operating, its temperature is 1.9 K, making the LHC the coldest place in the Solar System (excluding colder experiments).

FOUR SMASHING EXPERIMENTS

The machine is designed to collide protons into each other at energies of 14 TeV (14×10^{12} eV), travelling at 99.9999991% of the speed of light. These conditions are like those in the Universe 1 billionth of a second after the Big Bang. LHC scientists are hoping that this will then produce particles and interactions not seen since the Big Bang. In its first three years of operation, each collision had a total energy of up to 8 TeV. In 2015, collisions were made at energy levels of 13 TeV. Such high energies can probe the structure of the nucleons themselves.

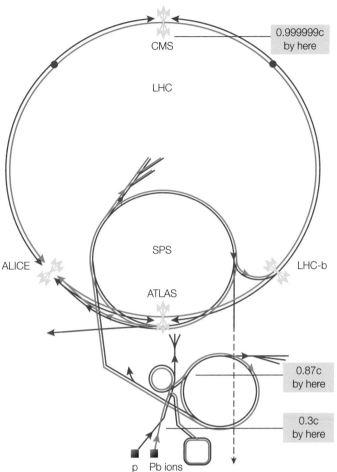

▲ **fig B** An overview of the Large Hadron Collider.

At such high energies, the beams of protons moving in opposite directions around the 27 km ring will cross each other's paths 30 million times every second, resulting in 600 million collisions per second. There are four critical experiments in the LHC, each named by an acronym: CMS, LHCb, ATLAS and ALICE. Each of these experiments has an incredibly complex detector built around the accelerator, and one of the beam-crossing points is at the centre of each detector. Each one is designed to detect particular products from the collisions. Each is searching in several different ways for specific undiscovered, but theoretical, particles.

The detectors include strong magnetic fields and will detect and show the movements of particles through the space that the detector occupies. Mass–energy and momentum are always conserved in particle interactions, together with charge. This means that the records of particle tracks can be interpreted to identify the particles in the detector and any reactions that happen.

THE COMPACT MUON SOLENOID (CMS)

One of the two teams that jointly announced discovery of the Higgs boson in July 2012 was from the CMS experiment (**fig C**). If the LHC achieves further exciting discoveries, they may come from this detector. There are several hypotheses in different areas of theoretical physics that may get confirmation evidence from the LHC. Some of these sound strange. For example, it is hoped that the CMS will observe mini black holes, dark matter, supersymmetric particles and gravitons.

The CMS is set up with various detecting chambers for different types of particle, and has 100 million individual detectors. These are in a 3D barrel containing as much iron as the Eiffel Tower in Paris. The charges and masses of particles can be calculated by monitoring the tracks of particles. The energies and momenta can also be measured and they must be conserved. This conservation means that all this information analysed together can identify all the particles and reactions in each collision.

THE LARGE HADRON COLLIDER BEAUTY EXPERIMENT (LHCB)

This detector looks out for the decays of both the bottom quark (sometimes called beauty) and the charm quark by looking for mesons containing these. This is to work out why our Universe contains mostly matter and very little anti-matter. Theoretically the two should appear in equal amounts.

A LARGE ION COLLIDER EXPERIMENT (ALICE)

Although ALICE initially observed the proton–proton collisions that the LHC started with, this detector is particularly intended to study the collisions of heavy ions, such as lead, accelerated to almost the speed of light. These collisions started in the second phase of LHC operation, and it is hoped they will create a quark-gluon plasma, which has been predicted by quantum mechanics theory.

ATLAS

The ATLAS detector (**fig D**) is 45 m long and 25 m high. It was the other experiment to provide experimental data verifying a new particle that is likely to be the Higgs boson. Among its potential further discoveries to come are extra dimensions of space, microscopic black holes, and evidence for dark matter particles in the Universe. Originally, ATLAS was an acronym for A Toroidal Lhc ApparatuS, but this has now been dropped and it is simply the name of the experiment.

▲ **fig D** The ATLAS experiment detector in the LHC.

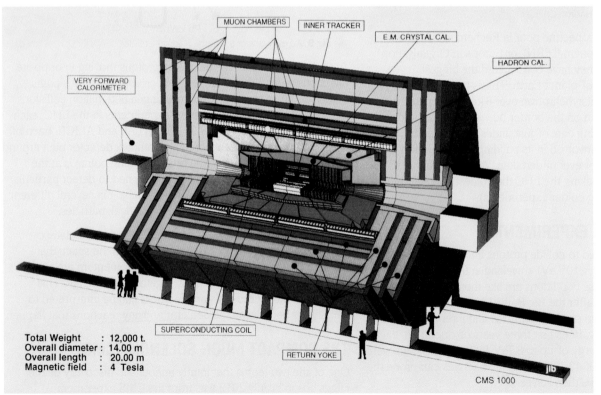

Total Weight : 12,000 t.
Overall diameter : 14.00 m
Overall length : 20.00 m
Magnetic field : 4 Tesla

▲ **fig C** Particle detection in the Compact Muon Solenoid (CMS).

WHAT MUST A DETECTOR BE CAPABLE OF DOING?

In developing the LHC experiments, the scientists had to work out what they needed the detectors to be able to do. The following nine points are listed on the ATLAS website as their aims.

1   *Measure* the directions, momenta and signs of charged particles.

2   *Measure* the energy carried by electrons and photons in each direction from the collision.

3   *Measure* the energy carried by hadrons (protons, pions, neutrons, etc.) in each direction.

4   *Identify* which charged particles from the collision, if any, are electrons.

5   *Identify* which charged particles from the collision, if any, are muons.

6   *Identify* whether some of the charged particles originate at points a few millimetres from the collision point rather than at the collision point itself (signaling a particle's decay a few millimetres from the collision point).

7   *Infer* (through momentum conservation) the presence of undetectable neutral particles such as neutrinos.

8   Have the *capability* to process the above information fast enough to permit flagging about 10–100 potentially interesting events per second out of the billion collisions per second that occur, and recording the measured information.

9   The detector must also be *capable* of long and reliable operation in a very hostile radiation environment.

> **DID YOU KNOW?**
>
> When it is running, the Large Hadron Collider contains half of the helium on Earth.

DATA ANALYSIS

In planning, CERN estimated the LHC would produce approximately 10% of the data produced through all human activities across the world. In 2017, the CERN data had stored 200 petabytes of data (2×10^{17} bytes). To analyse the data from the complex detectors, a system of computer analysis called the Grid was developed. This enables thousands of computers across the world to be linked together via the internet in order that their combined computing power can be used to study the experimental results and search out any which indicate the new discoveries it is hoped the LHC will produce. Of every 10 billion collision results, only about 10–100 are 'interesting' reactions. The ones that show things we already know need to be quickly filtered out of the data so that computing power is not wasted.

CHECKPOINT

1.   What can the curvature of a particle's track tell us about its charge and mass?

2.   How could momentum conservation help us to 'infer the presence of undetectable neutral particles such as neutrinos'?

SKILLS   ADAPTABILITY

3.   Why do we need four separate detector machines to undertake the various different particle searches at the LHC?

SKILLS   PRODUCTIVITY, ETHICS, COMMUNICATION, NEGOTIATION

4.   Do you think the cost of the LHC is justified? Explain your reasons for this answer.

EXAM HINT: EXTRA CONTENT

You are not expected to know any details about the Large Hadron Collider in the exam. You are expected to know the principles that make any accelerator or detector work.

MEDICAL PARTICLE ACCELERATORS

 SKILLS CRITICAL THINKING, PROBLEM SOLVING, INTERPRETATION, ADAPTIVE LEARNING, ADAPTABILITY, PRODUCTIVITY, COMMUNICATION

Particle accelerators can be used in medical applications. One of the most exciting possibilities is in cancer treatment, where a tumour is bombarded with high-energy particles, which can deliver energy to kill the cancerous cells. They can target a very small volume, so damage to surrounding tissue is less than with comparative treatments.

ONLINE ARTICLE

CERN INTENSIFIES MEDICAL PHYSICS RESEARCH

CERN has transferred a great deal of technology into the field of medical physics over the years, from adaptation of its high-energy particle detectors for PET scanning to the design and development of dedicated accelerators for particle therapy. Now, CERN is consolidating all of its diverse activities in the medical physics arena, with the creation of a new office for medical applications, to be headed up by Steve Myers.

The LEIR conversion

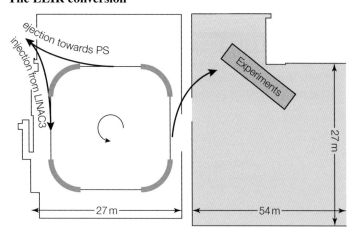

▲ **fig A** Redesign of the Low Energy Ion Ring at CERN for medical use.

One of the first projects that Myers will oversee is the transformation of CERN's Low Energy Ion Ring (LEIR) into a biomedical facility. LEIR is a small accelerator currently used to pre-accelerate lead ions for injection into the Large Hadron Collider (LHC). But it only does this for several weeks each year, leaving a lot of spare beam time. Importantly, the ring also has an energy range that matches that of medical accelerators (440 MeV/u for carbon ions).

CERN has now confirmed that LEIR can be converted into a dedicated facility for biomedical research. This 'BioLEIR' facility will provide particle beams of different types at various energies for use by external researchers. Conveniently, there's also a 54 × 27 m hall next to LEIR, currently used for storage, which could be developed into laboratory space.

Myers explained the rationale for developing such a facility. He pointed out that although protons and carbon ions are already in extensive clinical use for treating tumours, and other species such as oxygen and helium ions are under investigation, there's still a lack of controlled experiments that directly compare the effect of different ions on cancer cells under identical conditions.

And while existing clinical centres may well intend to perform such studies, once patient treatments begin, free accelerator time for research becomes extremely limited. 'The big advantage here is that we don't treat patients', said Myers. 'Our aim is to provide a service, so researchers don't have to do experiments at a clinical site, they can come here instead.'

As well as radiobiology experiments, LEIR is lined up for use in detector development, dosimetry studies and basic physics investigations such as ion beam fragmentation. The facility would work in the same way as particle physics experiments are carried out at CERN: researchers propose experiments, which are peer reviewed by a panel of experts who select suitable projects, and CERN controls the beam time allocation.

Before this can happen, however, LEIR needs some hardware modifications. These include a dedicated front-end that can accelerate many types of ion species, as well as a new extraction system and beamlines.

SCIENCE COMMUNICATION

The article is from medicalphysicsweb, which is a community website hosted by the Institute of Physics through their publishing division.

1 Explain why the Institute of Physics would spend time and money developing a website that is about medical matters.

2 Describe what level of fundamental research is in this article, and how that affects the style of the writing here.

3 (a) Describe the level of language used in the article, in particular the scientific terminology used.

   (b) Who do you think the intended audience is for this website, and how does that match with your answer to part (a)?

PHYSICS IN DETAIL

Now we will look at the physics in detail. Some of these questions will link to topics elsewhere in this book, so you may need to combine concepts from different areas of physics to work out the answers.

4 (a) Explain how the various ions that are mentioned would be accelerated in the LEIR.

   (b) What would be the difference in the motion of helium 2+ ions compared with oxygen 2+ ions, if each was accelerated through the same voltage in the LEIR?

5 (a) Explain why the LEIR was designed as a ring.

   (b) How can accelerating ions be made to continually travel around a ring of constant radius?

6 One of the most exciting aspects of this kind of cancer treatment is that, unlike X-ray or gamma ray treatment, the delivery of energy increases as the ion reaches the target tumour. As the particle slows down, it can interact with many more nearby particles, losing energy more quickly.

Explain how this could be better for the patient than the more uniform delivery of energy along the entire path in X-ray treatment.

ACTIVITY

Imagine a similar website aimed at a younger audience. Rewrite this article for a readership under 16 years old.

1 An electron moving at 2.05×10^4 km s^{-1} is made to move in a circular path by a 0.98 T magnetic field applied perpendicular to its velocity. The radius of the circle that the electron travels in is:

A 1.19×10^{-7} m

B 0.119 mm

C 8400 m

D 1.31×10^{26} m. [1]

(Total for Question 1 = 1 mark)

2 The diagram shows the track of a charged particle travelling in a bubble chamber detector, in which the magnetic field goes into the page.

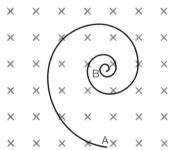

The particle:

A is negative and moves from A to B

B is positive and moves from A to B

C is negative and moves from B to A

D is positive and moves from B to A. [1]

(Total for Question 2 = 1 mark)

3 In order to investigate the structure of the nucleus, high energies are needed. One reason for this is that:

A The radius of circular motion of a particle in a cyclotron is proportional to the kinetic energy.

B The nucleus can only be ionised by high energy particles.

C The nucleus will absorb low energy particles.

D The de Broglie wavelength is inversely proportional to a particle's velocity. [1]

(Total for Question 3 = 1 mark)

4 A strong magnetic field of flux density B can be used to trap a positive ion by making it follow a circular orbit, as shown.

(a) Explain how the magnetic field maintains the ion in a circular orbit. You may add to a copy of the diagram above if you wish. [2]

(b) Show that the mass m of the ion will be given by

$$m = \frac{Bq}{2\pi f}$$

where q is the charge on the ion and f is the number of revolutions per second. [3]

(c) The above arrangement will not prevent a positive ion from moving vertically. To do this, a weak electric field is applied using the arrangement shown below.

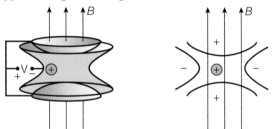

(i) Explain how the electric field prevents the ion moving vertically. [2]

(ii) This device is known as a Penning Trap. It can be used to determine the mass of an ion to an accuracy of 3 parts in 10 million.
Show that the mass of a sulfur ion can be measured to the nearest 0.00001 u.
mass of sulfur ion = 32.0645 u

'u' is an abbreviation for the atomic mass unit. 1 u is equivalent to 931.5 MeV. [2]

(iii) Under certain conditions, nuclei of sulfur emit a gamma ray with a known energy of 2.2 MeV. Calculate the resulting loss in mass of a sulfur ion in u and show that this value could be determined by the Penning Trap technique. [2]

(Total for Question 4 = 11 marks)

5 (a) A cyclotron can be used to accelerate charged particles.

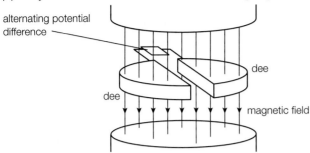

alternating potential difference

dee

dee

magnetic field

Explain the purpose of the magnetic field in the cyclotron. You may add to a copy of the diagram if you wish.   [2]

(b) A beam of low-speed protons are introduced into a cyclotron.

(i) Show that the number of revolutions per second, f, completed by the protons is given by

$$f = \frac{eB}{2\pi m}$$

where e is the electronic charge
        B is the uniform magnetic flux density within the cyclotron
        m is the mass of the proton.   [3]

(ii) An alternating potential difference is placed across the two dees to increase the energy of the protons. Explain why the potential difference that is used is alternating.   [2]

(iii) Initially, whilst the proton speeds are low, the frequency at which the potential difference has to alternate is constant. Explain how the frequency must change as the protons gain more and more energy.   [2]

(c) In the Large Hadron Collider at CERN, protons follow a circular path with speeds close to the speed of light. X-rays can be produced by free protons which are accelerating. Explain why this provides a source of X-rays, despite the speeds of the protons being constant.   [2]

(Total for Question 5 = 11 marks)

6 Explain the structure of linear particle accelerators and cyclotron accelerators, including the role of electric and magnetic fields in each.   [6]

(Total for Question 6 = 6 marks)

TOPIC 7 NUCLEAR AND PARTICLE PHYSICS

7C THE PARTICLE ZOO

On 4 July 2012, there was a press conference at CERN, the particle physics experiments lab near Geneva, when the discovery of a particle believed to be the Higgs boson was announced. This particle was predicted in the 1960s as the final component in the Standard Model. Physicists were excited by CERN's discovery, as this is experimental confirmation of the only piece of the Standard Model that had not yet been confirmed to exist. And its mass was very close to a theoretically predicted value. The scientific system of subatomic particles was completed. Experimental evidence of previous predictions is the foundation for the strongest scientific conclusions. This means that the Standard Model is a strong theory, and any corrections or alterations are now likely to be very minor.

In this chapter, we will show the Standard Model, so that you can see how the few particles you already know fit into the overall theory. These particles are the fundamental building blocks of our Universe. How they interact is what determines the development and structure of everything there is. Everything you will ever see is a consequence of the ideas in this chapter!

MATHS SKILLS FOR THIS CHAPTER

- **Recognising and use of appropriate units** (*e.g. converting joules and electronvolts*)

- **Use of standard form** (*e.g. using photon frequencies*)

- **Considering orders of magnitude** (*e.g. comparing the different generations of particles*)

- **Changing the subject of an equation** (*e.g. calculating photon frequencies*)

- **Substituting numerical values into equations using appropriate units** (*e.g. calculating matter conversions into energy*)

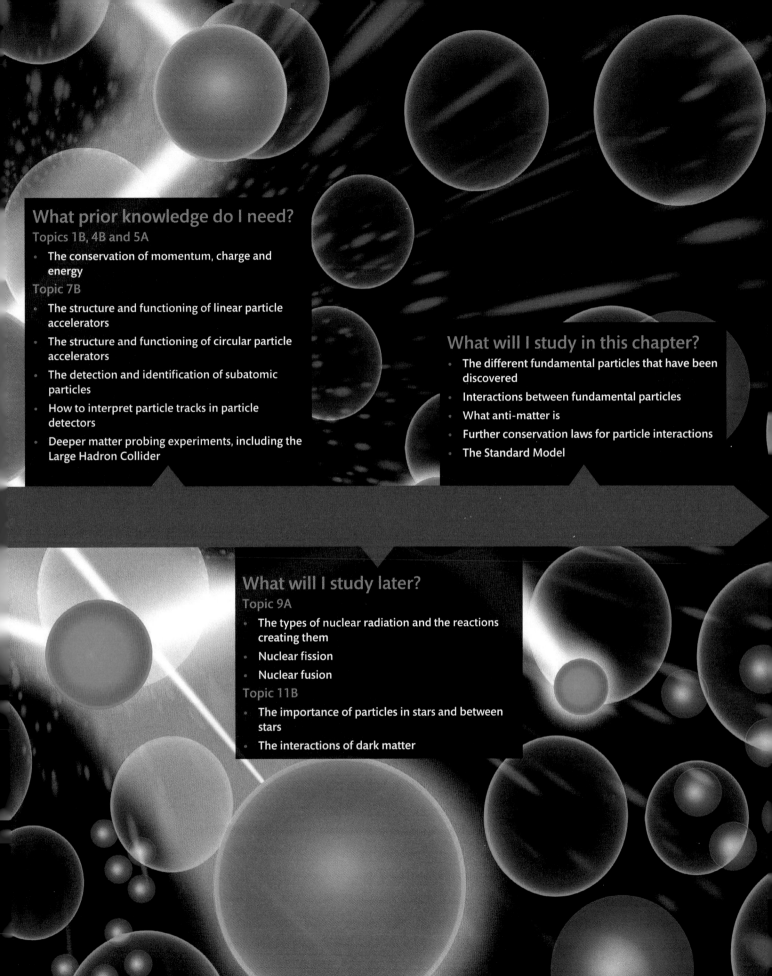

What prior knowledge do I need?

Topics 1B, 4B and 5A

- The conservation of momentum, charge and energy

Topic 7B

- The structure and functioning of linear particle accelerators
- The structure and functioning of circular particle accelerators
- The detection and identification of subatomic particles
- How to interpret particle tracks in particle detectors
- Deeper matter probing experiments, including the Large Hadron Collider

What will I study in this chapter?

- The different fundamental particles that have been discovered
- Interactions between fundamental particles
- What anti-matter is
- Further conservation laws for particle interactions
- The Standard Model

What will I study later?

Topic 9A

- The types of nuclear radiation and the reactions creating them
- Nuclear fission
- Nuclear fusion

Topic 11B

- The importance of particles in stars and between stars
- The interactions of dark matter

LEARNING OBJECTIVES

■ Use the equation $\Delta E = c^2 \Delta m$ in situations involving the creation and annihilation of matter and anti-matter particles.
■ Use and convert between MeV, GeV and $\dfrac{\text{MeV}}{c^2}$, $\dfrac{\text{GeV}}{c^2}$.

CREATION

The implication from most creation stories is that, from nothing, the material of the Earth was brought into being by an unexplained entity. However, one of Einstein's most important theories suggests exactly this – matter can appear where previously there was nothing but energy. Matter and energy are regularly interchanged in the Universe according to his well-known equation, $E = mc^2$. In this equation, multiplying the mass of an object by the square of the speed of light gives the equivalent amount of energy:

$$\Delta E = c^2 \times \Delta m$$

Given a suitable quantity of energy, such as that in a gamma ray photon, particles can spontaneously appear and the energy disappears from existence. This is so common in the Universe that it should not surprise us. The reason it does is that these events only happen on a sub-atomic scale, so we cannot detect them without complex machines.

WORKED EXAMPLE 1

A gamma ray photon converts into an electron and a positron (an anti-electron that has an identical mass to the electron). Calculate the frequency of the gamma photon.

Mass of an electron: $m_e = 9.11 \times 10^{-31}$ kg.

$$\Delta E = c^2 \times \Delta m = (3 \times 10^8)^2 \times 9.11 \times 10^{-31}$$

$$\Delta E = 8.2 \times 10^{-14}\,\text{J}$$

This is the amount of energy needed to produce an electron or a positron so, to produce both, the energy of the photon must be double this: 16.4×10^{-14} J.

$$E = hf$$

$$\therefore \quad f = \frac{E}{h} = \frac{16.4 \times 10^{-14}}{6.63 \times 10^{-34}}$$

$$f = 2.47 \times 10^{20}\,\text{Hz}$$

This reaction is known as electron–positron **pair production**. In momentum terms, it is just like an explosion. Initially only the photon existed so there was some linear momentum. Along this initial direction, their components of momentum must sum to the same total as the photon had. Perpendicular to the initial momentum, the electron and positron that were produced must have equal and opposite components of momentum so that in this direction it will still total zero afterwards.

In any reaction, the total combination of matter-energy must be conserved. If we add the energy equivalent of all matter particles with the energies, before and after the reaction, the numbers must be equal.

LEARNING TIP

All matter particles have an anti-matter equivalent. Anti-particles have the same mass, but all their other properties are opposite to those of the normal matter particle.

ANNIHILATION

Matter can appear spontaneously through a conversion from energy. In the same way, energy can appear through the disappearance of mass. This is the source of energy in nuclear fission and fusion. In both reactions, the sum of the masses of all matter involved before the reaction is greater than the sum of all mass afterwards. This mass difference is converted into energy (see **Section 9A.3**). In a nuclear power station we extract this energy as heat and use it to drive turbines to generate electricity.

If a particle and its anti-particle meet (anti-particles are the anti-matter versions of regular particles – see **Section 7C.2**), they will disappear to be replaced by the equivalent energy: we call this interaction **annihilation**. This reaction was supposedly the main power source to drive the starship *Enterprise* in the science fiction series *Star Trek* (**fig A**). Annihilation reactors could not be used as a power source on Earth, as anti-matter exists so rarely. Also, if we could find a supply of anti-matter, it would annihilate on contact with matter. This would most likely be before it reached the reaction chamber we had set up to use the energy for conversion into electricity.

▲ **fig A** The fictional starship *Enterprise* was said to be powered by a matter–anti-matter annihilation reaction.

ELECTRONVOLT UNITS

We have seen (in **Book 1 Section 4A.2**) the **electronvolt** (eV) as a unit for very small amounts of energy. Remember that one electronvolt is the amount of energy gained by an electron when it is accelerated through a potential difference of one volt. This is equivalent to 1.6×10^{-19} joules, so it is a very small amount of energy, even in particle physics terms. It is common for particles to have millions or even billions of electronvolts. For this reason we often use MeV and GeV as units of energy in particle interactions.

The atomic mass unit, u, is not an SI unit but is often used in particle interactions, as it is usually easier to understand. $1\,u = 1.66 \times 10^{-27}$ kg. As we know that energy and mass are connected by the equation $\Delta E = c^2 \Delta m$, we can also have mass units which are measures of E/c^2, such as MeV/c^2 and GeV/c^2. 1 u of mass is equivalent to about $931.5\,MeV/c^2$.

CHECKPOINT

SKILLS INTERPRETATION, ADAPTIVE LEARNING

1. A positron with kinetic energy 2.2 MeV collides with an electron at rest, and they annihilate each other.

    (a) Calculate the energy of each of the two identical gamma photons produced as a result of the annihilation.

    (b) Calculate the frequency of these gamma photons.

2. In 1977, in an accelerator at Fermilab, the upsilon particle was discovered. Its mass was determined as $9.46\,GeV/c^2$. What is the mass of the upsilon particle in kilograms?

3. A gamma ray photon travelling through space spontaneously converts into an electron and a positron.

    (a) Explain why the two new particles must move in opposite directions to each other.

    (b) A scientist observes a similar event in a detector and the incoming photon had a frequency of 4.89×10^{20} Hz. He sees that the tracks of the electron and positron from the reaction are 160° apart and hypothesises that a third particle must be produced. Calculate the mass of this third particle in MeV/c^2.

4. ▶ Look at **fig B**, which shows tracks in a hydrogen bubble chamber. One anti-proton hits a proton in the chamber. The resulting tracks show the tracks of various pairs of particles (π^+/π^-, e^+/e^-) which are created in the collision.

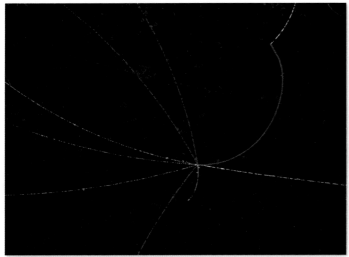

▲ **fig B** Tracks in a hydrogen bubble chamber that has been bombarded with anti-protons.

    (a) Identify the incoming anti-proton track.

    (b) How did you identify the incoming anti-proton? What does this tell you about its kinetic energy compared to the resulting particles?

    (c) How does the picture illustrate the conservation of matter/energy?

    (d) Identify any pair of particles (e.g., e^+/e^-). How do you know that the tracks show you a pair of identical but opposite particles?

    (e) How does the picture illustrate conservation of charge?

    (f) How does the picture illustrate the conservation of momentum in two dimensions?

WORKED EXAMPLE 2

Calculate the mass in kilograms of $1\,MeV/c^2$.

$$1\,MeV = 1 \times 10^6 \times 1.6 \times 10^{-19}$$
$$= 1.6 \times 10^{-13}\,J$$

In SI units, $c = 3 \times 10^8\,m\,s^{-1}$

$$1\,MeV/c^2 = \frac{1.6 \times 10^{-13}}{(3 \times 10^8)^2}$$

$$1\,MeV/c^2 = 1.78 \times 10^{-30}\,kg$$

This is about twice the mass of an electron.

LEARNING TIP

No particle interactions can occur if they break any of these conservation rules:

- momentum
- mass–energy
- charge.

There are also other rules that must be obeyed, but these three are critical, as all particles involved will have some of each property.

SUBJECT VOCABULARY

pair production the phenomenon in which a particle and its anti-matter equivalent are both created simultaneously in a conversion from energy

annihilation the phenomenon in which a particle and its anti-matter equivalent are both destroyed simultaneously in a conversion into energy which is carried away by force carrier particles, such as photons

electronvolt the amount of energy an electron gains by passing through a voltage of 1 V; $1\,eV = 1.6 \times 10^{-19}\,J$, 1 mega electronvolt = $1\,MeV = 1.6 \times 10^{-13}\,J$

LEARNING OBJECTIVES

- Define leptons and quarks in the Standard Model.
- Describe how the symmetry of the Standard Model predicted the existence of top quark.
- Explain that every particle has an anti-particle.
- Deduce the properties of particles and their anti-particles.

THE STANDARD MODEL

After a century in which scientists rapidly discovered many sub-atomic particles, they have now established a theory for how these come together to construct the things we see around us. This theory is known as the **Standard Model**. The idea of the atom as indivisible was swept aside by Rutherford and Thomson, and the idea of the proton and neutron as being fundamental has also been overturned. We have probed inside these two nucleons and discovered that each is constructed from smaller particles known as **quarks**. The electron is still considered fundamental. However, it has two partners – similar types of particles – and each has a neutrino associated with it, forming a group of six **fundamental particles**, known as **leptons**.

Scientists like to see the Universe as balanced, or symmetrical, and we have now found that quarks are also a group of six, although only the lightest two are found in protons and neutrons. The heavier quarks are found in rarer particles. The two groups are different because quarks can undergo interactions via the strong nuclear force, but leptons do not experience the strong force. It is the strong nuclear force that binds nucleons together in the nucleus. In each group of six, there are three pairs with a similar order of magnitude for mass, and these are known as the three generations of matter (**fig A**).

GENERATION	NAME	SYMBOL	CHARGE	MASS (MeV/c^2)
I	electron	e^-	−1	0.511
I	electron neutrino	v_e	0	0 ($<2.2 \times 10^{-6}$)
II	muon	μ	−1	106
II	muon neutrino	v_μ	0	0 (<0.17)
III	tau	τ	−1	1780
III	tau neutrino	v_τ	0	0 (<20)

table A The family of leptons. These do not experience the strong nuclear force.

EXAM HINT: EXTRA CONTENT

The strong nuclear force will not be in your exams.

GENERATION	NAME	SYMBOL	CHARGE	MASS (GeV/c^2)
I	up	u	$+\frac{2}{3}$	0.0023
I	down	d	$-\frac{1}{3}$	0.0048
II	strange	s	$-\frac{1}{3}$	0.095
II	charm	c	$+\frac{2}{3}$	1.275
III	bottom	b	$-\frac{1}{3}$	4.19
III	top	t	$+\frac{2}{3}$	173

table B The family of quarks. These are subject to the strong nuclear force.

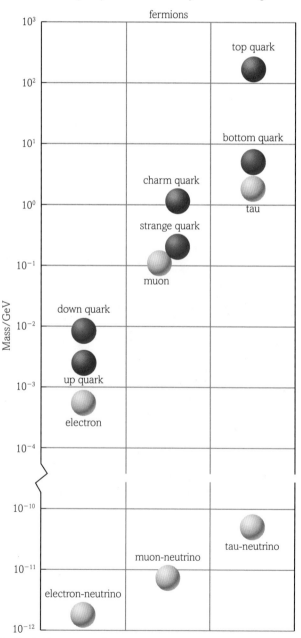

▲ **fig A** The masses of particles in the Standard Model increase over generations.

Only the first three quarks – up, down and strange – were known from the middle part of the twentieth century, and the charm quark was discovered in 1974. The symmetry of the Standard Model indicated to scientists that there were other particles they had never observed, which should exist – the bottom and top quark. Experiments were carried out to find these and the accelerator experiments at Fermilab discovered bottom in 1977 and top in 1995.

Current particle theory believes that all matter in the Universe is constructed from combinations of some of these 12 particles, and no others. Each of the 12 particles listed in **tables A** and **B** also has an anti-particle. The anti-particles have the same mass but all their other properties are opposite to those of the normal matter particle. To write the symbol for an **anti-particle**, it is the same as the normal particle, with a bar above the symbol. For example, the charge on an anti-down quark is $+\frac{1}{3}$ and its symbol is $\bar{d}$. In a few cases, the anti-particle notation treats it as a different particle in its own right and the bar may not be used. The positron (anti-electron) does not use a bar but is written as e^+.

> ### DID YOU KNOW?
> **THE HIGGS BOSON**
> One of the aims for CERN's Large Hadron Collider (LHC) experiment was that it would discover the Higgs boson. This was theoretically suggested independently by both Peter Higgs and the team of Francois Englert and Robert Brout, in 1964. Itself very massive, the Higgs particle is the means by which particles get mass. Remember that one of the key points about mass is that it gives objects inertia – the reluctance to change velocity. So an electron suffers less of this effect from the Higgs field than a proton, and so it has less mass. The Higgs mechanism (how this field imparts mass) is not easy to describe in everyday terms.
>
> On 4 July 2012, two LHC experiment teams, ATLAS and CMS, jointly announced that they had discovered a particle which matched the predicted interactions of the Higgs boson, and the mass of this new particle was approximately 125 GeV/c^2. Whilst the LHC scientists were reluctant to confirm it as the Higgs boson at the time, the new particle discovery was good enough for Higgs and Englert to be jointly awarded the 2013 Nobel Prize in Physics. Brout had unfortunately died a year before the experimental discovery was confirmed.

CHECKPOINT

1. An anti-electron is called a positron. Why is it usually written with the symbol e^+?

2. What is the difference between:
   (a) a quark and a lepton?
   (b) a quark and an anti-quark?
   (c) an electron and a muon?

3. ▷ (a) Having discovered up, down, strange, charm and bottom quarks, why did scientists undertake similar experiments using higher and higher energies?
   (b) The properties of the top quark that they discovered are shown in **table B**. What properties would an anti-top quark have?

4. Look at **fig A** and consider the three generations of neutrino, compared with the other fundamental particles in the Standard Model.
   (a) Why are the masses of the neutrinos surprising?
   (b) How do the neutrinos' masses follow the same trends as the other particles in the three generations?

SUBJECT VOCABULARY

Standard Model the name given to the theory of all the **fundamental particles** and how they interact. This is the theory that currently has the strongest experimental evidence

quarks the six fundamental particles that interact with each other using the strong nuclear force (as well as all other forces)

fundamental particles the most basic particles that are not made from anything smaller. These can be combined to create larger particles

leptons the six fundamental particles which do not interact using the strong nuclear force, only the other three fundamental forces

anti-particle has the same mass but all their other properties are opposite to those of the normal matter particle

LEARNING OBJECTIVES

■ Define baryons, mesons and photons in the Standard Model.

■ Explain why high energies are required to investigate fundamental particles.

BARYONS

If three quarks are combined together, the resulting particle is a **baryon** (**fig A**). Protons and neutrons are baryons. A proton consists of two up quarks and a down quark, whilst a neutron is two down quarks combined with an up quark. Other baryons are more obscure and have very short lives, as they decay through a strong nuclear force reaction that makes them highly unstable. Yet other baryons, like the sigma, omega and lambda particles, decay via the weak nuclear force and are longer lived, with lives as long as 10^{-10} seconds!

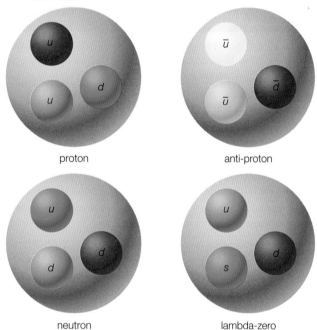

proton

anti-proton

neutron

lambda-zero

▲ **fig A** Three-quark baryons. Three anti-quarks in the anti-proton make it an anti-baryon.

MESONS

If a quark and an anti-quark are combined together, the resulting particle is known as a **meson** (**fig B**). The **pion** and the **kaon** are the most common examples of mesons. A π^+ meson ($u\bar{d}$) consists of an up quark (u) combined with an anti-down ($\bar{d}$) quark. If a meson is a combination of a quark and its anti-quark, then the meson's charge must be zero. This is the case for the J/ψ particle, which was first discovered in 1974 by two independent researchers at separate laboratories. Its slow decay, with a lifetime of 7.2×10^{-21} s, occurred when it was not travelling at relativistic

speeds, so was a genuine and unexpected long lifetime. This did not fit the pattern generated by up, down and strange quarks, and it is thought to be a charm–anti-charm combination ($c\bar{c}$). The two independent discoverers of this, Richter and Ting, shared the 1976 Nobel Prize for their discovery of the fourth quark.

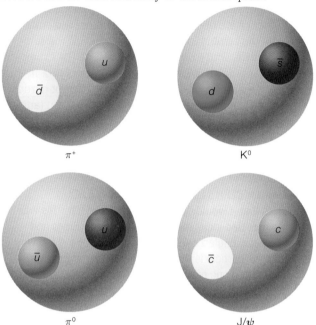

π^+

K^0

π^0

J/ψ

▲ **fig B** Mesons are formed from quark–anti-quark combinations.

HADRONS

Quarks can interact via the strong nuclear force. Thus baryons and mesons can interact via the strong nuclear force. Any particle which experiences the strong force is called a **hadron**. So, baryons and mesons are both hadrons. Leptons do not experience the strong force and so are in a separate class of particle from the hadrons.

THE FOUR FORCES OF NATURE

There are four other particles that we have not yet mentioned here, because they are not matter particles. These are known together as **exchange bosons**. The matter particles interact by the four forces of nature, which are **gravity**, the **electromagnetic force**, the strong nuclear force and the **weak nuclear force**. Each force acts on particles which have a specific property, such as mass in the case of gravity, or electric charge for the electromagnetic force. The process by which these forces act has been modelled by scientists as an exchange of another type of particles – the exchange bosons. For example, for a proton and an electron to attract each other's opposite charge, they pass **photons** backwards and forwards between each other.

EXAM HINT: EXTRA CONTENT

The exam specification does not include all the exchange bosons, but you do have to know about photons.

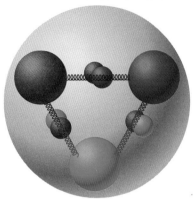

▲ **fig C** Gluons 'in combination with virtual quark anti-quark pairs' are exchanged between quarks to hold the quarks together. In this example, the three quarks form a proton.

This has been shown experimentally to be an appropriate model for the electromagnetic, strong and weak forces. In the case of gravity, the so-called **graviton** has been theoretically invented to complete the model, but gravitons are yet to be discovered. Many experiments have been set up recently to try and detect gravitons.

FORCE	EXCHANGE BOSON	BOSON SYMBOL	BOSON CHARGE	BOSON MASS (GeV/c^2)
electromagnetic	photon	γ	0	0
weak nuclear	W or Z boson	W^-	−1	80.4
		W^+	+1	80.4
		Z^0	0	91.2
strong nuclear	gluon	g	0	0
gravity	graviton	undetermined		

table A The four forces of nature and their exchange bosons.

EVIDENCE FOR EXCHANGE BOSONS

Gravity is the force that we experience most obviously, but it is the one scientists know the least detail about. The electromagnetic force is well understood, and the exchange of photons to make it work has also led to an understanding of the generation of photons such as light. The weak nuclear force is also fairly well understood, as the W and Z bosons that were predicted by theory were later detected in accelerator experiments in which protons are crashed together. This is particularly strong science, as the existence of the weak nuclear force bosons had been theoretically predicted and then they were detected in later experiments. Many particle decays where other particles are formed, such as beta decay, occur via a weak nuclear force interaction.

CHECKPOINT

SKILLS ADAPTABILITY

1. What is the difference in the quark composition of a proton and a neutron? Explain what you think happens to the quarks when a neutron undergoes beta minus decay.

2. What is the difference between:
   (a) a meson and a baryon?
   (b) a hadron and a lepton?

3. What is an exchange boson, like the photon?

4. Describe what happens to make a proton repel another proton.

SUBJECT VOCABULARY

baryon a particle made of a combination of three quarks

meson a particle made of a combination of a quark and an anti-quark

pion a meson created from any combination of up and down quark/anti-quark pairings

kaon a meson created from any combination of an up or down quark/anti-quark and a strange or anti-strange quark

hadron a particle which can interact via the strong nuclear force

exchange bosons particles that enable the transfer of force. Each of the four fundamental forces has its own exchange boson

gravity the weakest of the four fundamental forces, affecting all objects

electromagnetic force one of the four fundamental forces, transmitted by photons, acting between objects with charges

weak nuclear force one of the four fundamental forces, transmitted by W or Z bosons, acting at extremely short ranges (10^{-18} m); it can affect all matter particles*

photons the quantum of electromagnetic radiation, and force carrier for the electromagnetic force

graviton the force carrier particle (or exchange boson) for gravity*

* You do not need to revise these terms for your exam.

EXAM HINT: EXTRA CONTENT

W and Z bosons, gluons and gravitons will not be in your exams. The strong and weak nuclear forces are also beyond the exam requirements.

LEARNING OBJECTIVES

■ Use the laws of conservation of charge, baryon number and lepton number to determine whether a particle interaction is possible.
■ Write and interpret particle equations given the relevant particle symbols.
■ Describe situations in which the relativistic increase in particle lifetime is significant.

REACTIONS CONSERVE PROPERTIES

For any particle reaction to occur, the overall reaction must conserve various properties of the particles involved. The total combination of mass/energy must be the same before and after the reaction. Furthermore, momentum and charge must also be conserved. Momentum and mass/energy are difficult to check. In a reaction, particles can begin or end with more kinetic energy to balance any apparent mass difference. This is how accelerator collision experiments can create large mass particles: particles with high kinetic energies can have the energy converted into mass to generate many particles with less energy. Reactions can also have similar flexibility to ensure momentum conservation. We will see later that particles are also assigned values called baryon number and lepton number, and these must also be conserved in reactions.

CHARGE CONSERVATION

Charge must be conserved for any particle interaction to be possible. We can quickly see if charge is conserved by checking the reaction's equation. Consider two common nuclear reactions: **alpha decay** and **beta-minus decay** (**fig A**).

alpha decay $^{235}_{92}U \Rightarrow {}^{231}_{90}Th + {}^{4}_{2}\alpha$

charge: $+92 \Rightarrow +90 + +2$ ✓

beta-minus decay $^{14}_{6}C \Rightarrow {}^{14}_{7}N + {}^{0}_{-1}\beta + {}^{0}_{0}\bar{\upsilon}_e$

charge: $+6 \Rightarrow +7 + -1 + 0$ ✓

The reaction for beta-minus decay led to the development of the theory that neutrinos and anti-neutrinos exist. They are almost massless and have no charge, so are almost impossible to detect.

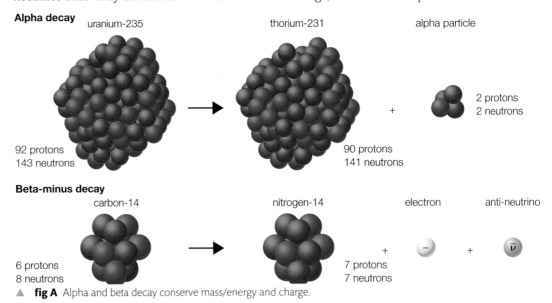

Alpha decay

uranium-235   thorium-231   alpha particle

92 protons   90 protons   2 protons
143 neutrons   141 neutrons   2 neutrons

Beta-minus decay

carbon-14   nitrogen-14   electron   anti-neutrino

6 protons   7 protons
8 neutrons   7 neutrons

▲ **fig A** Alpha and beta decay conserve mass/energy and charge.

If the same nuclear change produces the same single particle every time, then for mass–energy to be conserved, the beta particles would need to have the same energy every time. This is the case for alpha particles. However, scientists found that beta particles from nuclei of the same isotope have a range of kinetic energies. This suggested that another particle was flying away with some kinetic energy, so that the total kinetic energy produced was always the same.

WORKED EXAMPLE 1

For each of these reactions, confirm whether or not they can occur through conservation of charge:

(a) Beta-plus decay:

$$^{13}_{7}N \Rightarrow {}^{13}_{6}C + {}^{0}_{+1}\beta^{+} + {}^{0}_{0}\upsilon_e$$

Answer: $+7 \Rightarrow +6 + +1 + 0$ ✓

This reaction is permitted as it conserves charge. It would need to also conserve mass/energy and momentum.

(b) Positive pion decay:

$$\pi^{+} \Rightarrow \mu^{-} + {}^{0}_{0}\upsilon_\mu$$

Answer: $+1 \Rightarrow -1 + 0$ ✗

This reaction is not permitted as it does not conserve charge.

(NB: In fact, a positive pion decays into an anti-muon, μ^+, and a muon neutrino with a lifetime of some 26 nanoseconds.)

LEARNING TIP

The products from beta decay processes include either an electron (beta minus) or a positron (beta plus). This means that you may see either symbol for each used in beta decay equations:

$${}^{0}_{-1}\beta^{-} \qquad {}^{0}_{+1}\beta^{+}$$

CONSERVATION OF BARYON AND LEPTON NUMBERS

We can also check on the possibility of a reaction occurring, as the reaction must conserve baryon number and lepton number. Each quark has a **baryon number**, B, of $+\frac{1}{3}$, and so a baryon has a value of $B = +1$. Each lepton has a **lepton number**, L, of $+1$. Anti-particles have the opposite number. As mesons are quark/anti-quark combinations, their total baryon number is zero $(+\frac{1}{3} + -\frac{1}{3} = 0)$.

Particle reactions can also only occur if they conserve baryon and lepton numbers overall. This means the total for each property must be the same before and after any reaction, or else it cannot occur.

Let us revisit alpha and beta decay and check on conservation of these numbers.

alpha decay $\quad {}^{235}_{92}U \Rightarrow {}^{231}_{90}Th + {}^{4}_{2}\alpha$

charge:	$+92 \Rightarrow +90 + +2$	✓
baryon number:	$+235 \Rightarrow +231 + +4$	✓
lepton number:	$0 \Rightarrow 0 + 0$	✓

beta-minus decay $\quad {}^{14}_{6}C \Rightarrow {}^{14}_{7}N + {}^{0}_{-1}\beta + {}^{0}_{0}\overline{\upsilon}_e$

charge:	$+6 \Rightarrow +7 + -1 + 0$	✓
baryon number:	$+14 \Rightarrow +14 + 0 + 0$	✓
lepton number:	$0 \Rightarrow 0 + +1 + -1$	✓

If we isolate the individual neutron decay reaction in beta decay, we can check again:

beta-minus decay $\quad {}^{1}_{0}n \Rightarrow {}^{1}_{1}p + {}^{0}_{-1}\beta + {}^{0}_{0}\overline{\upsilon}_e$

charge:	$0 \Rightarrow +1 + -1 + 0$	✓
baryon number:	$+1 \Rightarrow +1 + 0 + 0$	✓
lepton number:	$0 \Rightarrow 0 + +1 + -1$	✓

PARTICLE	SYMBOL	CHARGE	LEPTON NUMBER, L	BARYON NUMBER, B	ANTI-PARTICLE	SYMBOL	CHARGE	LEPTON NUMBER, L	BARYON NUMBER, B
electron	e^-	-1	1	0	positron	e^+	$+1$	-1	0
electron neutrino	v_e	0	1	0	anti-electron neutrino	$\overline{v}_e$	0	-1	0
muon	μ	-1	1	0	anti-muon	$\overline{\mu}$	$+1$	-1	0
muon neutrino	v_μ	0	1	0	anti-muon neutrino	$\overline{v}_\mu$	0	-1	0
tau	τ	-1	1	0	anti-tau	$\overline{\tau}$	$+1$	-1	0
tau neutrino	v_τ	0	1	0	anti-tau neutrino	$\overline{v}_\tau$	0	-1	0

table A Leptons and anti-leptons and their properties.

PARTICLE	SYMBOL	CHARGE	LEPTON NUMBER, L	BARYON NUMBER, B	ANTI-PARTICLE	SYMBOL	CHARGE	LEPTON NUMBER, L	BARYON NUMBER, B
proton	p	$+1$	0	$+1$	anti-proton	$\overline{p}$	$+1$	0	-1
neutron	n	0	0	$+1$	anti-neutron	$\overline{n}$	0	0	-1
neutral pion	π^0	0	0	0	neutral pion	$\overline{\pi}^0$	0	0	0
pi-plus	π^+	$+1$	0	0	anti-pi-plus	$\overline{\pi}^+$	0	0	0
down quark	d	$-\frac{1}{3}$	0	$+\frac{1}{3}$	anti-down	$\overline{d}$	$+\frac{1}{3}$	0	$-\frac{1}{3}$
xi-minus	Ξ^-	-1	0	$+1$	xi-plus	Ξ^+	$+1$	0	-1

table B Various particles and anti-particles, and their properties.

This provides further evidence that this reaction produces an anti-neutrino because the reaction must have an anti-lepton on the right-hand side to balance the electron.

If we zoom in further and isolate the individual quark decay reaction in beta decay, we can check again:

beta-minus decay	$d \Rightarrow u + {}^{0}_{-1}\beta + {}^{0}_{0}\bar{v}_e$
charge:	$-\frac{1}{3} \Rightarrow +\frac{2}{3} - 1 + 0$ ✓
baryon number:	$+\frac{1}{3} \Rightarrow +\frac{1}{3} + 0 + 0$ ✓
lepton number:	$0 \Rightarrow 0 + +1 + -1$ ✓

Again, everything is conserved.

STRANGENESS

The strange quark adds an additional property to reactions, which must usually also be conserved. This is called **strangeness**, S. Each strange quark has a strangeness of -1, each anti-strange quark has $S = +1$, and all other particles have zero strangeness. Strong and electromagnetic force interactions must conserve strangeness.

WORKED EXAMPLE 2

The lambda-plus particle is a baryon and consists of two up quarks and a strange quark (uus). It commonly decays via the strong force into a proton (uud) and a neutral pion (which can be $u\bar{u}$ or $d\bar{d}$). True or false?

proposed decay	${}^{1}_{1}\Lambda \Rightarrow {}^{1}_{1}p + {}^{0}_{0}\pi$
charge:	$+1 \Rightarrow +1 + 0$ ✓
baryon number:	$+1 \Rightarrow +1 + 0$ ✓
lepton number:	$0 \Rightarrow 0 + 0$ ✓
strangeness:	$-1 \Rightarrow 0 + 0$ ✗

This reaction cannot occur via a strong force as it does not conserve strangeness.

RELATIVISTIC LIFETIMES

We have seen previously (**Section 7B.1**) that when a particle is accelerated close to the speed of light, its mass appears to increase. Einstein also predicted that, at these speeds, time would be slower than it is for an external observer. This suggests that when we monitor very fast-moving particles, their lifetimes before they decay are longer than the theoretical predictions.

One early piece of evidence of these extended particle lifetimes came with the detection of muons at sea level. High energy cosmic rays interact with the nuclei of atoms high in the atmosphere to create muons, which have a lifetime of about $2\,\mu s$. These should all decay in much less than the time to travel down to the Earth's surface, and we should detect very few at sea level. However, they are detected at low altitudes in large numbers. The reason is that they are travelling very fast (for example $0.98c$). As far as the muons are concerned, they still decay in the same short time; but for us as external observers, their time moves slowly enough for them to travel tens of kilometres down to our experiments on the ground. This phenomenon was first confirmed experimentally by Rossi and Hall in 1940.

This principle also affects the particle interactions observed in accelerator collision experiments. The particles live longer, and travel further, than we might expect because they are moving so fast. This allows many additional interactions to happen because they meet particles they might not otherwise have had time to come into contact with. Also, the fast-moving particles that will travel much further through the detectors than non-relativistic lifetimes would allow, and so can be more easily observed. If the exotic particles produced in the LHC collision chamber decayed within the chamber before escaping it, none would be detectable, as the detector cannot be placed right at the collision point.

CHECKPOINT

> **SKILLS**   ANALYSIS, PROBLEM SOLVING

1. For each of these reactions, confirm whether or not they can occur considering only conservation of charge:

   (a) phosphorus beta-plus decay.

   The phosphorus isotope is converted into a sulfur isotope and emits a positron and an electron neutrino:

   $$^{32}_{15}P \Rightarrow \, ^{32}_{16}S + \, ^{0}_{+1}\beta + \, ^{0}_{0}v_e$$

   (b) neutral pion decay.

   A neutral pion becomes an electron, a positron and a gamma photon:

   $$\pi^0 \Rightarrow e^- + e^+ + \gamma$$

2. ▶ A muon is generated in the atmosphere at a height of 10 km. Calculate how far it would travel down towards the ground if it was moving at $0.98\,c$ and it decays after 2.2×10^{-6} s.

3. Beta plus decay is similar to the better known beta-minus decay. In beta-plus decay, a proton is converted into a neutron, with the emission of a positron and an electron neutrino.

   Write the fundamental reaction for beta-plus decay and confirm that it is permitted by the conservation rules you have seen.

4. ▶ One theoretical physicist at CERN has proposed that the LHC may see a neutron decay reaction in which a lambda zero particle (up, down, strange baryon) and a kaon (down/anti-strange meson) are produced:

   proposed reaction 1 $^{1}_{0}n \Rightarrow \, ^{1}_{0}\Lambda + \, ^{0}_{0}K$

   A second theoretical physicist at CERN has argued that the reaction will have to produce two kaons as well as a positron and an anti-neutrino:

   proposed reaction 2 $^{1}_{0}n \Rightarrow \, ^{1}_{0}\Lambda + 2\,^{0}_{0}K + \, ^{0}_{1}e^+ + \, ^{0}_{0}\overline{v}_e$

   Work out which reaction, if any, is possible.

SUBJECT VOCABULARY

alpha decay the radioactive process in which a particle combination of two protons and two neutrons is ejected from a nucleus

beta-minus decay the radioactive process in which a nuclear neutron changes into a proton, and an electron is ejected from the nucleus

baryon number the quantum number for baryons, whereby each proton or neutron (or other baryon) has a value of $B = 1$

lepton number the quantum number for leptons, whereby each lepton has a value of $L = 1$

strangeness the quantum number for strange quarks, whereby each one has a value of $S = -1$

ANTI-MATTER MATTERS

 SKILLS CRITICAL THINKING, PROBLEM SOLVING, ANALYSIS, INTERPRETATION, ADAPTIVE LEARNING, CONTINUOUS LEARNING, INITIATIVE, PRODUCTIVITY

Anti-matter was first proposed by Paul Dirac in 1928. It is very interesting for both science fiction writers and scientists because it is such a strange idea. Here, NASA's website explains the progress that has been made towards using the energy of anti-matter.

NASA WEBSITE

STATUS OF ANTI-MATTER

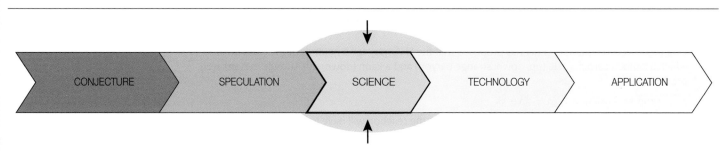

fig A The maturity of a scientific concept changes over time.

Anti-matter is **real stuff**, not just science fiction. Anti-matter is firmly in the realm of science with some aspects even entering the technology realm. There is also a lot of speculation about what one might do with anti-matter.

What is anti-matter?

Anti-matter is matter with its electrical charge reversed. Anti-electrons, called 'positrons', are like an electron but with a positive charge. Antiprotons are like protons with a negative charge. Positrons, antiprotons and other antiparticles can be routinely created at particle accelerator labs, such as CERN in Europe, and can even be trapped and stored for days or weeks at a time. And just last year, they made antihydrogen for the first time. It didn't last long, but they did it. Also, anti-matter is **NOT** antigravity. Although it has not been experimentally confirmed, existing theory predicts that anti-matter behaves the same to gravity as does normal matter.

Technology is now being explored to make anti-matter carrying cases, to consider using anti-matter for medical purposes, and to consider how to make anti-matter rockets.

The catch?

Right now it would cost about one-hundred-billion dollars to create one milligram of anti-matter. One milligram is way beyond what is needed for research purposes, but that amount would be needed for large scale applications. To be commercially viable, this price would have to drop by about a factor of ten-thousand.

And what about using anti-matter for power generation? – not promising.

It costs far more energy to create anti-matter than the energy one could get back from an anti-matter reaction. Right now standard nuclear reactors, which take advantage of the decay of radioactive substances, are far more promising as power generating technology than anti-matter. Something to keep in mind, too, is that anti-matter reactions – where anti-matter and normal matter collide and release energy, require the same safety precautions as needed with nuclear reactions.

Extract from the NASA website, at *http://www.nasa.gov/centers/glenn/technology/warp/antistat.html* posted on 2 May 2008. The article has since been updated with a focus on warp-drive technology.

SCIENCE COMMUNICATION

1 (a) What sort of people do you think might visit this webpage?

    (b) How has the author ensured that the level of scientific terminology, and the depth of explanations, are right for their audience?

2 What is NASA's motivation for including this page on its website?

INTERPRETATION NOTE

Consider NASA's general purposes and aims. You may find it helpful to visit the original webpage and explore any links provided.

PHYSICS IN DETAIL

Now we will look at the physics in detail. Some of these questions will link to topics elsewhere in this book, so you may need to combine concepts from different areas of physics to work out the answers.

3 What would be the structure of anti-hydrogen?

4 (a) How could a particle physics lab trap, or store, a charged particle like a proton for a long time?

    (b) How would CERN store the positrons they create?

    (c) Why would it be much more difficult to store anti-hydrogen than an anti-proton?

5 (a) What are the differences between an electron and a positron?

    (b) What differences are there between an anti-baryon and an anti-lepton?

    (c) Why is it generally unnecessary to define a group of particles we would call anti-mesons?

6 Using the data given in the extract (and assuming ordinary matter is free) calculate the cost per joule of energy created if anti-matter annihilation were to be used for power generation.

7 (a) If a positron beam were to be used as an annihilation cutting tool, calculate the beam current that would deliver 100 kW.

    (b) What voltage would be needed to accelerate a beam of electrons so that, at the same beam current, it could deliver the same power from mechanical collisions with the target?

PHYSICS TIP

Think of the movement of positrons in a beam as an electric current, which undergo annihilation when interacting with the target.

ACTIVITY

The webpage was published in 2008 and the study of anti-matter is progressing quickly. Write an updated version of the webpage for your nearest university physics department to include on their site. The purpose of the webpage, is to explain anti-matter to students who may be considering applying to study physics there. You do not need to mention any activities that the department are involved with, the page should simply inspire and interest students into the possibilities that a career in physics could offer. The level of scientific explanation should be a little deeper than the NASA original shown.

THINKING BIGGER TIP

Your webpage should be fully up-to-date.

You will therefore need to do some further research in order to find out the latest developments and areas of research. For example, has the cost of making one milligram of anti-matter changed since 2008?

1 Which of these is a force exchange particle?

  A photon

  B muon

  C kaon

  D meson   [1]

(Total for Question 1 = 1 mark)

2 Which of the following is NOT a possible value for the charge on a meson?

  A −1

  B 0

  C $\frac{1}{3}$

  D 1   [1]

(Total for Question 2 = 1 mark)

3 Which of the following is a lepton?

  A A baryon

  B An electron neutrino

  C A neutral pi meson

  D A proton   [1]

(Total for Question 3 = 1 mark)

4 'Positron' is the name for an anti-electron. The properties of the positron are:

	Mass/kg	Lepton number	Charge
A	-9.11×10^{-31}	0	−1
B	-9.11×10^{-31}	+1	+1
C	9.11×10^{-31}	0	−1
D	9.11×10^{-31}	−1	+1

[1]

(Total for Question 4 = 1 mark)

5 Which of these is a baryon?

  A electron

  B kaon

  C meson

  D proton   [1]

(Total for Question 5 = 1 mark)

6 The diagram shows tracks produced by charged particles in a bubble chamber.

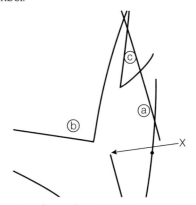

At X, an incoming charged particle interacts with a stationary proton to produce a neutral lambda particle and a neutral kaon particle. Both these particles later decay into other particles.

With reference to the diagram, explain the evidence provided for this event.   [4]

(Total for Question 6 = 4 marks)

7 In 1961 Murray Gell-Mann predicted the existence of a new particle called an omega (Ω) minus. It was subsequently discovered in 1964.

At this time the quark model consisted of three particles, the properties of which are given in the table.

Quark	Charge	Predicted mass in MeV/c^2
up (u)	$+\frac{2}{3}$	4
down (d)	$-\frac{1}{3}$	4
strange (s)	$-\frac{1}{3}$	80

(a) Explain what a charge of $+\frac{2}{3}$ means. [1]

(b) State the predicted mass of, and the charge on, an $\bar{s}$ particle. [2]

(c) Convert 4 MeV/c^2 to kg. [3]

(d) The event that led to the discovery of the omega minus particle can be summarised as follows. A negative kaon collided with a stationary proton and produced a positive kaon, a neutral kaon and the omega minus.

  (i) Kaons consist of either an up or down quark plus a strange quark. The omega minus consists of three strange quarks.
  Copy and complete the following table by ticking the appropriate boxes. [2]

	Meson	Baryon	Nucleon	Lepton
negative kaon				
omega minus				

  (ii) Write an equation using standard particle symbols to summarise this event. [2]

  (iii) The negative kaon consists of $\bar{u}s$. Deduce the quark structure of two other kaons involved in this event. [2]

  (iv) The total mass of the three particles created after this event is larger than the total mass of the two particles before. Discuss the quantities that must be conserved in interactions between particles and use an appropriate conservation law to explain this increase in mass. [5]

(Total for Question 7 = 17 marks)

8 (a) Physicists were able to confidently predict the existence of a sixth quark. State why. [1]

(b) The mass of the top quark was determined by an experiment. Collisions between protons and anti-protons occasionally produce two top quarks.

  (i) How do the properties of a proton and an anti-proton compare? [2]

  (ii) After the collision the two top quarks move in opposite directions with the same speed. Explain why. [2]

(c) The two top quarks decay rapidly into two muons and four jets of particles. These can be detected and their momenta measured.

The diagram shows an end-on view of the directions of the four jets (J1 to J4) of particles. The two muons are shown as μ_1 and μ_2. A muon neutrino is also produced but cannot be detected, so is **not** shown. Each momentum is measured in GeV/c.

The magnitude of the momentum for each particle or 'jet' is shown by the number printed at the end of each arrow.

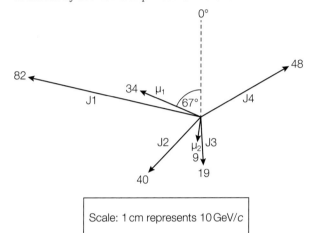

Scale: 1 cm represents 10 GeV/c

(i) Explain why GeV/c is a valid unit for momentum. [2]

(ii) The vector diagram shown below is **not** complete. Copy the diagram and add arrows to represent the momenta of J3 and J4. [2]

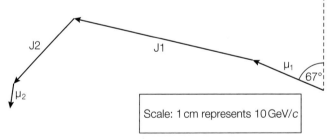

Scale: 1 cm represents 10 GeV/c

(iii) Complete the diagram to determine the magnitude of the momentum of the muon neutrino. [1]

(iv) Show that the total energy of all the products of this event is about 300 GeV. [1]

(v) Deduce the mass of a top quark in GeV/c^2. [1]

(vi) Explain why it took a long time to find experimental evidence for the top quark. [2]

(Total for Question 8 = 14 marks)

TOPIC 8 **THERMODYNAMICS**

8A HEAT AND TEMPERATURE

When scuba divers need to return to the surface of the water in an emergency, they are trained to scream all the way up. This may sound a little strange, but the training is designed to ensure that they do not have to think about expelling the gas in their lungs, as this happens automatically when they scream. As the pressure decreases on their way to the surface, the air in their lungs expands. This could be very dangerous to their internal organs, particularly their lungs, and so the air must be expelled at least as fast as it expands. This is illustrated in a similar, although less extreme, scenario: the bubbles from a diver's normal exhaled breath will slowly get larger as they rise to the surface.

This chapter is all about the energy held by individual molecules, and how, on a large scale, this energy is responsible for many macroscopic effects, such as the pressure exerted by a gas on a diver's lungs. We will learn about a large number of rules governing the properties of gases, including their pressure, temperature, volume and quantity. These will be explained from first principles in terms of the molecules that make up the gas. In a more general sense, we will see how the energy possessed by a group of molecules leads to the temperature of a sample and the transfer of heat energy.

MATHS SKILLS FOR THIS CHAPTER

- **Recognising and use of appropriate units** (*e.g. converting between temperature scales*)
- **Finding arithmetic means** (*e.g. mean square speeds of gas molecules*)
- **Use of ratios** (*e.g. illustrating the gas laws*)
- **Estimating results** (*e.g. predicting the outcome on a gas of changes in its properties*)
- **Substituting numerical values into equations** (*e.g. calculations using the ideal gas equation*)

What prior knowledge do I need?

- **The pressure on a solid**

Topic 1B (Book 1: IAS)

- **The kinetic energy equation**

Topics 1B and 1C (Book 1: IAS)

- **The conservation of momentum and energy**
- **Collisions and Newton's laws of motion**

Topic 2A (Book 1: IAS)

- **Density**

Topic 4B (Book 1: IAS)

- **The potential divider**

Topic 5A

- **Impulse and momentum change**

What will I study in this chapter?

- The difference between heat and temperature
- Scales of temperature
- The ideal gas equation
- Basic gas laws
- Internal energy and the distribution of energy through a gas sample

What will I study later?

Topic 9A

- Nuclear fission energy release

Topic 10A

- Simple harmonic motion
- Resonance and damping in molecular energy transfer

Topic 11B

- The importance of heat pressure in stars and stellar development
- The emission of stellar radiation following black body curves

LEARNING OBJECTIVES

■ Define the concept of *absolute zero* and explain how the average kinetic energy of molecules is related to the absolute temperature.

■ Identify the differences between scales of temperature measurement.

■ Explain how a thermistor can be calibrated in a potential divider circuit to act as a thermostat.

TEMPERATURE

Heating an object causes its temperature to rise. This seems a very obvious statement, based on our everyday experience, but what is really happening when we heat an object, and what do we mean by 'temperature'?

To understand the concept of temperature, we need to think about materials in terms of the molecules from which they are made. According to the **kinetic theory**, when energy is supplied to an object the molecules in that object receive the energy as kinetic energy, and move faster. In solids, this is usually in the form of vibrations, whereas if we are considering a gas, we imagine the molecules whizzing around their container at a greater speed. It is this kinetic energy that determines the temperature. If the average kinetic energy of the molecules of a substance is greater, then it is at a higher temperature.

It is important to remember that the internal kinetic energy on the atomic scale is separate from the idea of overall movement of the object. For example, as a cricket ball flies through the air, it has kinetic energy due to its overall movement. In addition to this, its molecules will be vibrating within it. It is this internal aspect of the energy that determines the temperature.

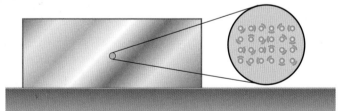

▲ **fig A** A stationary piece of metal has no kinetic energy. Yet all its atoms are vibrating all the time. Although overall these tiny velocities must sum to zero (or else it would be moving), their kinetic energies give the object its temperature.

ABSOLUTE TEMPERATURE

Taking energy away from the molecules of a substance causes its temperature to become lower. If you imagine a situation in which energy is continuously being taken away from a collection of molecules, then there will come a moment when all the kinetic energy has been removed from the substance. The molecules will no longer be moving at all. At this point, the temperature is said to be **absolute zero**.

Different scales of temperature have historically placed zero in arbitrary places compared with the average internal kinetic energy of the molecules. For example, Anders Celsius chose the freezing point of pure water as the zero on his temperature scale, but Daniel Gabriel Fahrenheit placed zero at 32 degrees below the freezing point of water on the Fahrenheit scale (in the expectation that there would never be a temperature colder than 0 °F to measure!). On these scales, absolute zero is at −273.15 °C and −459.67 °F. In 1848, Lord Kelvin defined an absolute temperature scale which started with zero at absolute zero. For convenience, he made the gaps in his scale identical to those on the **Celsius scale**. This is known as the **Kelvin scale** of temperature, or sometimes as **absolute temperature**. As the units begin from a definite zero and go up as we add even amounts of energy to the particles of the substance being measured, the various temperatures are not compared to something else (such as the freezing point of water). This is why the term *absolute* temperature is used. We do not use the word or symbol for 'degrees' when quoting values on the Kelvin scale. For example, the temperature at which lead becomes superconducting is 7.19 kelvin, or 7.19 K.

The Scottish engineer William Rankine proposed a similar absolute temperature scale, with degrees separated by the same amount as in the Fahrenheit scale. The Rankine scale is not used today,

▲ **fig B** William Thomson, who was made Baron Kelvin of Largs in 1892, was one of Britain's most prolific and important scientists. He is buried next to Isaac Newton in Westminster Abbey.

TEMPERATURE SCALE	Celsius	Fahrenheit	Kelvin	Rankine
SYMBOL	°C	°F	K	R
BOILING POINT OF WATER (AT 1 ATM = 101 325 PA)	100	212	373.15	671.67
FREEZING POINT OF WATER (AT 1 ATM = 101 325 PA)	0	32	273.15	491.67
ABSOLUTE ZERO	–273.15	–459.67	0	0

table A Comparing four scales of temperature.

LEARNING TIP

Converting degrees celsius to kelvin, add 273.15.

Converting kelvin to degrees celsius, subtract 273.15.

You can remember this by thinking that C is nearer the beginning of the alphabet to K. Therefore, to go back from K to °C, we need to subtract.

PRACTICAL SKILLS   CP12

Investigating thermostats

In **Book 1, Section 4B.3**, we saw how a potential divider circuit can be used with a sensor component to control a second circuit. This can be useful, for example, if we want to control a heating or cooling circuit, such as with the circuit shown in **fig C**.

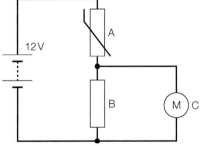

▲ **fig C** Investigate how the motor (C) speed adjusts as the thermistor (A) temperature is altered. This circuit could be used to control a cooling fan.

Many electronic components have a switch on voltage of 5 V. For example, the control circuit to switch on a large air conditioner could be used with a thermistor in a potential divider circuit and would only switch on when the p.d. across it is 5 V. Altering the value of the other resistor in the potential divider circuit would then control the temperature level at which the air conditioner was switched on.

You can use a circuit like that shown in **fig C** to find the output voltage at different temperatures to calibrate the thermistor. This calibration should then allow you to replace the fixed resistor with a variable resistor that you can set, so that the output exceeds 5 volts only when a particular temperature is reached.

Safety Note: Be careful not to overload the components in the circuit as they may get hot enough to burn skin.

EXAM HINT

Make sure you have a good understanding of this practical as your understanding of the experimental method may be assessed in your exams.

EXAM HINT: EXTRA CONTENT

The Fahrenheit and Rankine scales of temperature are not on the exam specification.

CHECKPOINT

SKILLS ▸ CONTINUOUS LEARNING, INNOVATION, ADAPTABILITY

1. Convert the following temperatures:
   (a) 27 °C into kelvin
   (b) –173 °C into kelvin
   (c) 100 °C into kelvin
   (d) 288 K into degrees celsius.
2. ▸ Explain what will happen in the circuit in **fig C** as the temperature rises.
3. ▸ What changes need to be made to the circuit in **fig C** in order to allow the thermistor to be calibrated for use as a thermostat?

SUBJECT VOCABULARY

kinetic theory the idea that consideration of the microscopic movements of particles will predict the macroscopic behaviour of a substance

absolute zero the theoretical temperature at which molecules will no longer be moving, all kinetic energy has been removed

Celsius scale a scale of temperature with zero degrees at the freezing point of water, and 100 degrees at the boiling point of water

Kelvin scale an absolute temperature scale with each degree the same size as those on the Celsius scale

absolute temperature a temperature scale that starts at absolute zero

LEARNING OBJECTIVES

■ Define the concepts of *specific heat capacity and specific latent heat* for phase changes.
■ Use the equations $\Delta E = mc\Delta\theta$ and $\Delta E = L\Delta m$.

We have seen that the temperature of a substance depends on the kinetic energy of the molecules within it. As the molecules move, their bonds can pull on neighbouring molecules. This means that lattice vibrations in the solid can pass energy through the solid.

Imagine that one end of a long rod is being heated. In the hotter parts of the metal, the molecules are vibrating faster and further. In the cooler parts of the rod, further from the heat source, the molecules are moving slower. When a molecule with more energy pulls on one with less energy, they share the energy more evenly, with the faster one slowing down and the slower one speeding up. The effect of these transfers is that the increase in the internal kinetic energy which is caused by heating moves throughout a substance, with the heat passing from hotter areas to colder ones.

SPECIFIC HEAT CAPACITY

Transferring the same amount of heat energy to two different objects will increase their internal energy by the same amount. However, this will not necessarily cause the same rise in temperature in both. How transferred heat energy affects the temperature of an object depends on three things:

1 the amount of heat energy transferred
2 the mass of the object
3 the specific heat capacity of the material from which the object is made.

How much the temperature rises depends on the material, and is given by a property known as its **specific heat capacity**, *c*. Different materials, and different phases of the same substance, have different specific heat capacities because their molecular structures are different. This means that their molecules will be affected to different degrees by additional heat energy.

For a certain amount of energy, ΔE, transferred to a material, the change in temperature, $\Delta\theta$, is related to the mass of material, *m*, and the specific heat capacity, *c*, by the expression:

$$\Delta E = mc\Delta\theta$$

With each quantity measured in SI units, this means that the specific heat capacity has units of $J\,kg^{-1}\,K^{-1}$. The *change* in temperature, $\Delta\theta$, is the same whether measured in degrees celsius or kelvin, as the intervals are the same on both scales.

WORKED EXAMPLE 1

How long will a 2 kW kettle take to raise the temperature of 800 grams of water from 20 °C to 100 °C? The specific heat capacity of water is $4200\,J\,kg^{-1}\,K^{-1}$.

$$\Delta\theta = 100 - 20 = 80\,°C = 80\,K$$
$$\Delta E = mc\,\Delta\theta = 0.8 \times 4200 \times 80 = 268\,800\,J$$
$$t = \frac{E}{P} = \frac{268\,800}{2000} = 134.4\,s$$

Therefore the kettle takes 2 minutes and 14 seconds to heat this water to boiling temperature. We assume that no energy is wasted in heating the surroundings.

PRACTICAL SKILLS

Investigating specific heat capacity

It is quite straightforward to measure the specific heat capacity of most materials (some values are given in **table A**). You can make a close measurement if you can measure the heat energy put into a known mass of the material and if you can insulate the material so that almost all of this heat is used to raise the temperature. To gain very accurate measurements requires significantly more complex insulation. The inaccuracy in this rough experiment can usually be kept to under 5% with careful experimentation.

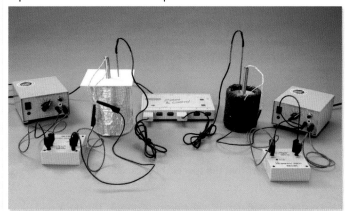

▲ **fig A** Investigating specific heat capacity experimentally.

SUBSTANCE	SPECIFIC HEAT CAPACITY/J kg^{-1} K^{-1}
water	4200
ice	2100
steam	1996
wood	1670
air (50 °C)	1020
aluminium	900
marble	860
iron/steel	450
copper	390
silver	240
mercury	140

table A Some typical specific heat capacities.

Safety Note: The heaters and metal blocks being tested will get hot enough to burn skin. The immersion heaters must never have their tops under water as steam may be produced inside the heater which makes it explode.

SPECIFIC LATENT HEAT

To melt a solid that is already at the temperature at which it melts requires an input of additional energy. This energy input does not raise the temperature, but is only used in breaking the bonds between molecules to change the state of the material. The amount of additional energy needed to melt it is determined by two things:

1 the mass of the solid substance

2 the specific latent heat of fusion of the substance.

How much energy is needed to alter the molecular configuration depends on the material, and is given by a property known as its **specific latent heat**, L. Different materials, and different phase changes of the same substance, have different specific latent heat values because their molecular structures are different. This means that their molecules will be affected differently by additional heat energy.

LEARNING TIP

For melting, L is known as the specific latent heat of fusion.

For boiling, L is known as the specific latent heat of vaporisation.

For a certain mass of material, m, the amount of additional energy, ΔE, needed to make the phase change is related to the specific latent heat, L, by the expression:

$$\Delta E = L\Delta m$$

With each quantity measured in SI units, this means that the specific latent heat has units of J kg^{-1}. The change in temperature is zero for any phase change, so the specific heat capacity need not be considered if the temperature is constant and only the state of matter changes.

WORKED EXAMPLE 2

How long will a 150 W freezer take to freeze 650 grams of water which is already at 0 °C? The specific latent heat of fusion of water is 334 000 J kg^{-1}.

$$\Delta E = L\Delta m = 334\,000 \times 0.65 = 217\,100\,\text{J}$$

$$t = \frac{E}{P} = \frac{217\,100}{150} = 1447\,\text{s}$$

Therefore the freezer takes a little over 24 minutes to freeze this water. We assume that no energy is gained from the surroundings.

PRACTICAL SKILLS   CP13

Investigating specific latent heat of fusion

It is quite straightforward to measure the specific latent heat of fusion of water. You can make a close measurement if you can measure the heat energy put into a known mass of ice, and if you set up the experiment well enough that almost all of the heat goes towards melting the ice. A control funnel of crushed ice can also be used in which the heater is off, so we can see how much ice would have melted without the extra heating. The difference in mass of water collected between the heated funnel and the control gives us the mass to use in the calculation.

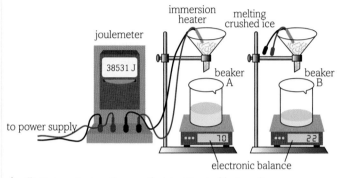

▲ **fig B** Investigating the specific latent heat of fusion of ice.

Safety Note: The heaters and metal blocks being tested will get hot enough to burn skin. The immersion heaters must never have their tops under water as steam may be produced inside the heater which makes it explode.

EXAM HINT

Make sure you have a good understanding of this practical as your understanding of the experimental method may be assessed in your exams.

LEARNING TIP

To change state, molecules gain potential energy, as this is associated with the forces between the molecules. As the potential energy is negative, increasing it means it gets closer to zero.

CHECKPOINT

1. (a) How much heat energy must be supplied to raise the temperature of 1200 ml of water from 20 °C to 65 °C?

    (b) How long would a 2200 W heater take to boil 3 litres of water initially at 100 °C?

      (1 litre of water has a mass of 1 kilogram. Specific latent heat of vaporisation of water = 2260 kJ kg⁻¹.)

    SKILLS   ANALYSIS

2. ▶ Heating up half a kilogram of ice at −8 °C until it is water at 8 °C requires 192 200 joules of energy. Show the separate calculations necessary to determine this value.

    SKILLS   ADAPTIVE LEARNING

3. ▶ **Table A** shows that the specific heat capacity of copper is 390 J kg⁻¹ K⁻¹.

    A student read part of a chemistry textbook. The book said that copper at 1085 °C had been supplied with 4000 J of energy, but its temperature remained at 1085 °C. Explain how this could be possible.

SUBJECT VOCABULARY

specific heat capacity the energy required to raise the temperature of one kilogram of a substance by one degree kelvin

specific latent heat the energy required to change the state of one kilogram of a substance at a constant temperature

The average kinetic energy of the particles in a material gives the material its temperature. However, in addition to this kinetic energy, each molecule will have some potential energy which is as a result of its position within the structure of the material, or in relation to other molecules in the substance. This potential energy is due to the bonds between the molecules. If we sum the kinetic and potential energies of all the molecules within a given mass of a substance, we have measured its **internal energy**.

It is important to note that the molecules do not all have the same amount of kinetic and potential energies. This internal energy is randomly distributed across all the molecules, according to the **Maxwell–Boltzmann distribution**.

THE MAXWELL–BOLTZMANN DISTRIBUTION

If we identify the individual velocity of each molecule in a particular sample, the values will range from a few moving very slowly to a few moving very fast, with the majority moving at close to the average speed. As they all have the same mass, their differences in kinetic energies are directly dependent on the speeds. If we plot the kinetic energy against the number of molecules that have that energy, we obtain a curved graph called the Maxwell–Boltzmann distribution, as in **fig A**.

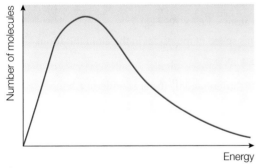

▲ **fig A** The Maxwell–Boltzmann distribution of the speeds of a mixture of particles at one particular temperature.

The characteristic shape of the graph in **fig A** shows that:
• there are no molecules with zero energy
• only a few molecules have high energy
• there is no maximum value for the energy a molecule can have.

A Maxwell–Boltzmann distribution graph is for one specific temperature. As the temperature changes, so the graph changes. The peak on the graph moves towards higher energies (and therefore higher speeds) as the temperature increases.

ROOT-MEAN-SQUARE CALCULATIONS

There are two different ways of determining the average speeds of molecules in a material which are of interest to physicists. The peak of the graph represents the kinetic energy value with the greatest number of molecules that have that energy. Therefore the speed corresponding to this kinetic energy is the most probable speed, c_0, if we were to select a molecule at random. The second and more useful average is the **root-mean-square speed**, which has the symbol $\sqrt{\langle c^2\rangle}$. This is the speed associated with the average kinetic energy, $\frac{1}{2}m\langle c^2\rangle$, where c is the speed of the particle, and m is its mass.

The root-mean-square speed, often abbreviated to rms speed, is found by squaring the individual speeds of a set of molecules, finding the mean of the squares, and then taking the square root of that mean.

WORKED EXAMPLE 1

Find the rms speed of the following group of five molecules of atmospheric nitrogen, as shown in **fig B**:

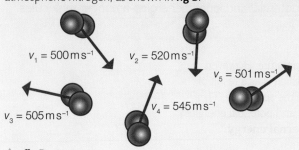

$v_1 = 500 \, \text{m s}^{-1}$ $v_2 = 520 \, \text{m s}^{-1}$ $v_5 = 501 \, \text{m s}^{-1}$ $v_3 = 505 \, \text{m s}^{-1}$ $v_4 = 545 \, \text{m s}^{-1}$

▲ **fig B**

First find the square of each value:

$v_1^2 = 250\,000$; $v_2^2 = 270\,400$; $v_3^2 = 255\,025$; $v_4^2 = 297\,025$; $v_5^2 = 251\,001$

The mean of the square values is:

$$<c^2> = \frac{(v_1^2 + v_2^2 + v_3^2 + v_4^2 + v_5^2)}{5}$$
$$= \frac{(250\,000 + 270\,400 + 255\,025 + 297\,025 + 251\,001)}{5}$$
$$= \frac{1\,323\,451}{5}$$
$$= 264\,690.2$$

Finally, take the square root to get the rms speed:

$$\sqrt{<c^2>} = \sqrt{264\,690.2} = 514 \, \text{m s}^{-1}$$

MOLECULAR KINETIC ENERGY

The average kinetic energy of any molecule in a gaseous sample is proportional to the absolute temperature of the gas. They are related by the expression:

$$\tfrac{1}{2}m<c^2> = \tfrac{3}{2}kT$$

where k is the Boltzmann constant, $1.38 \times 10^{-23} \, \text{J K}^{-1}$. Here the unit of temperature, T, is absolute temperature and must be in kelvin.

From this equation, relating kinetic energy and absolute temperature, we can see that zero on the absolute scale of temperature must indicate a situation where the molecules are stationary. Their mass cannot change, so for the kinetic energy to be zero at absolute zero, their rms speed must decrease to zero.

EXAM HINT

The idea that the molecules' average kinetic energy is proportional to the absolute temperature is an assumption within kinetic theory. This follows consistently from the ideal gas laws (see **Section 8A.4**), but these are also founded on a set of assumptions which we must accept before undertaking calculations using the equation:

$$\tfrac{1}{2}m<c^2> = \tfrac{3}{2}kT$$

WORKED EXAMPLE 2

What is the rms speed of helium molecules in a child's balloon at 20 °C?

(Atomic mass of helium, $m = 6.65 \times 10^{-27} \, \text{kg}$)

$T = 20 + 273 = 293 \, \text{K}$

$\tfrac{1}{2}m<c^2> = \tfrac{3}{2}kT$
$\qquad = \tfrac{3}{2} \times 1.38 \times 10^{-23} \times 293$
$\qquad = 6.0651 \times 10^{-21} \, \text{J}$

$$<c^2> = \frac{\tfrac{3}{2}kT}{\tfrac{1}{2}m}$$
$$= \frac{6.0651 \times 10^{-21}}{\tfrac{1}{2}(6.65 \times 10^{-27})}$$
$$= 1\,820\,000$$

rms speed $= \sqrt{<c^2>}$
$\qquad = \sqrt{1\,820\,000}$
$\qquad = 1350 \, \text{m s}^{-1}$

The rms speed of helium molecules at 20 °C is $1350 \, \text{m s}^{-1}$.

CHECKPOINT

1. (a) What is the rms speed of four hydrogen molecules with speeds of 890, 755, 902 and 866 m s^{-1}?

   (b) At what temperature would these hydrogen molecules be?

   (Mass of H atom = 1.67×10^{-27} kg)

2. Estimate the rms speed of molecules in the air in this room.

   SKILLS PROBLEM SOLVING

3. ▶ What would be represented by the area underneath the Maxwell–Boltzmann curve?

   SKILLS INTERPRETATION

4. ▶ Copy a sketch graph of the Maxwell–Boltzmann distribution labelling the x-axis 'speed'.

   (a) Mark a vertical line on your graph to show the energy corresponding to the most probable speed of molecules. Label this line c_0.

   (b) From Maxwell–Boltzmann statistics, the most probable speed (c_0), the mean speed ($<c>$), and the rms speed ($\sqrt{<c^2>}$) are in the ratio:

   1 : 1.13 : 1.23

   Mark two more vertical lines on your graph and label these $<c>$ and $\sqrt{<c^2>}$.

SUBJECT VOCABULARY

internal energy the sum of the kinetic and potential energies of all the molecules within a given mass of a substance

Maxwell–Boltzmann distribution a mathematical function that describes the distribution of energies amongst particles at a given temperature

root-mean-square speed the square root of the arithmetic mean value of the squares of the speeds of particles in an ideal gas

LEARNING OBJECTIVES

■ Define the concept of an ideal gas and explain the relationships between its pressure, temperature and volume.
■ Use the equation $pV = NkT$ for an ideal gas.
■ Derive and use the equation $\frac{1}{2} m\langle c^2\rangle = \frac{3}{2} kT$.

BOYLE'S LAW

During the last 400 years, scientists investigating the physics of gases have determined a number of laws governing their behaviour. The first of these was discovered by Robert Boyle, an Irish physicist working at Oxford University. **Boyle's law** states that:

For a constant mass of gas at a constant temperature, the pressure exerted by the gas is inversely proportional to the volume it occupies.

If you imagine lying on an air mattress, after you sink slightly, it provides a force to hold you up. Your body squashes the volume of the mattress, making it smaller. This increases the pressure exerted by the gas inside, until the force applied by the gas pressure equals your weight. This is Boyle's law in action.

PRACTICAL SKILLS   CP14

Investigating Boyle's law

Boyle's law can be demonstrated using the apparatus shown in **fig A**. Measurements of the length of air which is trapped in the vertical glass column represent the volume of the gas, and the pressure is measured using the barometer. A graph of pressure against 1/volume will give a straight best-fit line indicating that:

$$p \propto \frac{1}{V}$$

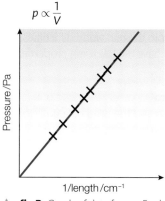

▲ **fig A** Apparatus to demonstrate Boyle's law.

▲ **fig B** Graph of data from a Boyle's law experiment.

! Safety Note: Wear eye protection and make sure all the connections are secure to avoid high pressure leaks. Reduce the pressure before disconnecting the pump.

EXAM HINT

Make sure you have a good understanding of this practical as your understanding of the experimental method may be assessed in your exams.

CHARLES'S LAW

About 140 years after Boyle published his law, the French scientist Joseph Louis Gay-Lussac published two further gas laws. One was based on unpublished data from experiments undertaken by Jacques Charles. **Charles's law** states that:

For a constant mass of gas at a constant pressure, the volume occupied by the gas is proportional to its absolute temperature.

This law can be shown in symbols as:

$$V \propto T$$

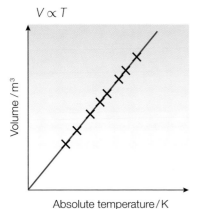

▲ **fig C** Graph of data from Charles's law experiment.

THE PRESSURE LAW

Gay-Lussac's other gas law sometimes bears his name, and sometimes is simply referred to as the **pressure law**:

For a constant mass of gas at a constant volume, the pressure exerted by the gas is proportional to its absolute temperature.

This law can be shown in symbols as:

$$p \propto T$$

PRACTICAL SKILLS

Investigating the pressure law

▲ **fig D** Apparatus to demonstrate the pressure law. The temperature and pressure readings could be taken by electronic sensors and the data logged by an attached computer.

!

Safety Note: Wear eye protection and do not let the water level get too low in the beaker.

The pressure law can be demonstrated using the apparatus shown in **fig D**. Measurements of the gas pressure and temperature can be datalogged to produce a graph of pressure against absolute temperature, which will give a straight best-fit line indicating that:

$$p \propto T$$

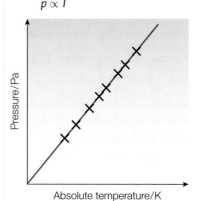

▲ **fig E** Graph of data from pressure law experiment.

IDEAL GASES

The three gas laws above were worked out from graphs of experimental results which showed distinct, straight best-fit lines, allowing scientists to claim them as empirical laws (laws worked out from experimental data). However, if very accurate experiments are undertaken with a variety of different gases, we find that the laws are not perfectly accurate. For example, Charles's law suggests that if we reduce the temperature to zero kelvin, the gas has zero volume – it would disappear. This does not happen, as the gas volume cannot reduce to less than the combined volume of its molecules.

The three gas laws are idealised, and would work perfectly if we could find a gas that did not suffer from the real world difficulties that real gases have. An **ideal gas** would have the following properties:

1 The molecules have negligible size.
2 The molecules are identical.
3 All collisions are perfectly elastic and the time of a collision is significantly smaller than the time between collisions.
4 The molecules exert no forces on each other, except during collisions.
5 There are enough molecules so that statistics can be applied.
6 The motion of the molecules is random.

Assuming an ideal gas, we can combine the three gas laws to produce a single equation relating the pressure, volume, temperature and amount of a gas:

$$pV = NkT$$

where N is the number of molecules of the gas, and k is the Boltzmann constant. The temperature must be absolute temperature in kelvin.

This is known as the **equation of state** for an ideal gas, expressed in terms of the number of molecules present. Sometimes, a more practically useful version can be used which refers to large quantities of gases, as we may find in a balloon, for example. If we change from the Boltzmann constant, k, to using the gas constant for entire moles:

$$pV = nRT$$

where n is the number of moles of the gas, and R is the universal gas constant, $R = 8.31 \, \text{J kg}^{-1} \text{mol}^{-1}$.

THE MOLE

The **mole** (abbreviated to mol) is an SI unit used to measure the amount of a substance. One mole of a substance is defined to consist of Avogadro's number of molecules of that substance. N_A is the symbol for Avogadro's number: 6.02×10^{23}.

For one mole of an ideal gas:

$$pV = RT$$

Comparing this with the equation of state:

$$pV = N_A kT$$

$$\therefore \quad N_A \times k = R$$

$$6.02 \times 10^{23} \times 1.38 \times 10^{-23} = 8.31$$

WORKED EXAMPLE 1

What would be the pressure in a child's helium balloon at 20 °C, if it is a sphere of radius 20 cm, containing three moles of helium gas?

$$V = \tfrac{4}{3}\pi r^3 = \tfrac{4}{3}\pi(0.2)^3 = 3.35 \times 10^{-2} \, \text{m}^3$$

$$N = 3 \times N_A$$

$$= 3 \times 6.02 \times 10^{23}$$

$$= 18.06 \times 10^{23}$$

$$p = \frac{NkT}{V}$$

$$= \frac{18.06 \times 10^{23} \times 1.38 \times 10^{-23} \times 293}{3.35 \times 10^{-2}}$$

$$= 217\,981 \, \text{Pa}$$

$$= 218\,000 \, \text{Pa (3 sf)}$$

Or use $pV = nRT$:

$$p = \frac{nRT}{V}$$

$$= \frac{3 \times 8.31 \times 293}{3.35 \times 10^{-2}}$$

$$= 218\,000 \, \text{Pa}$$

AVERAGE MOLECULAR KINETIC ENERGY

The ideal gas equation is:

$$pV = NkT$$

and there's an alternative equation from the kinetic theory of gases:

$$pV = \tfrac{1}{3} Nm\langle c^2 \rangle$$

Putting these together:

$$\tfrac{1}{3} Nm\langle c^2 \rangle = NkT$$

$$\therefore \quad \tfrac{1}{3} m\langle c^2 \rangle = kT$$

Multiply both sides by $\tfrac{3}{2}$:

$$\tfrac{3}{2}\tfrac{1}{3} m\langle c^2 \rangle = \tfrac{3}{2} kT$$

$$\therefore \quad \tfrac{1}{2} m\langle c^2 \rangle = \tfrac{3}{2} kT$$

We saw the above equation in use in **Section 8A.3**, and this derivation shows that it is a consistent consequence of the kinetic theory of gases.

EXAM HINT

The command word **Devise** means that you have to plan or invent a procedure based on the physics and experimental principles that you know.

SUBJECT VOCABULARY

Boyle's law for a constant mass of gas at a constant temperature, the pressure exerted by the gas is inversely proportional to the volume it occupies

Charles's law for a constant mass of gas at a constant pressure, the volume occupied by the gas is proportional to its absolute temperature

pressure law for a constant mass of gas at a constant volume, the pressure exerted by the gas is proportional to its absolute temperature

ideal gas a theoretical gas which does not suffer from the real world difficulties that mean real gases do not perfectly follow the gas laws

equation of state the single equation that defines a gas in terms of its pressure, volume, temperature and quantity: $pV = NkT$

mole the SI unit for amount of substance. One mole contains 6.02×10^{23} molecules of that substance

REACH FOR THE SKY!

In 2015, two pilots made the second Pacific crossing in a helium-filled balloon and achieved new world records for distance and duration.

NATIONAL GEOGRAPHIC NEWS

RECORD-SETTING BALLOONISTS LAND OFF BAJA COAST

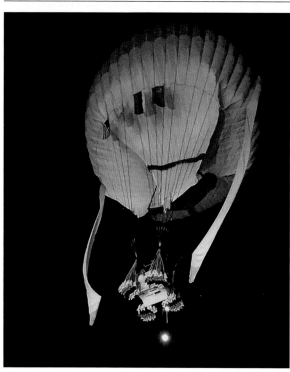

▲ **fig A** The Two Eagles capsule, where the pilots spend most of their time, is about as big as a king-size bed.

The Two Eagles helium-filled balloon took off from Saga, Japan, on January 25 2015 and landed safely about a week later off the coast of the Baja Peninsula, Mexico. It was the second vessel of its kind to successfully cross the Pacific Ocean.

"They didn't just break the records," said Steven Shope, director of the Two Eagles mission control in Albuquerque, New Mexico. "They shattered the distance record by 27 percent and the duration record by 17 percent. Those previous records had stood for over 30 years."

Moving at the speed and caprice of wind, it took them 160 hours and 37 minutes to cover the 10 712 kilometres.

Prior to the Two Eagles' flight, the distance record for a gas-filled balloon was 8 383 kilometres, set in 1981 during the first trans-Pacific balloon flight. Gas-filled balloons are distinct from hot-air balloons, which have never made that trip.

What is the difference between a hot-air balloon and a gas balloon?

Balloons, whether filled with gas (typically helium), hot air, or a mixture of gas and air, operate only by their capability to gain or lose altitude. Otherwise, it's up to the wind to propel the balloon forward. Charting a course and reaching a destination in a balloon is a challenge that brings together aeronautical science and adventure.

Both types of balloons rely on buoyancy to gain altitude. That buoyancy is created by filling the "envelope" of the balloon with a substance that is less dense than the surrounding air. Helium and hydrogen, for instance, are naturally less dense than our atmosphere. When heated, air loses density because it expands and therefore contains less mass per volume.

When did people start ballooning?

Joseph-Michel Montgolfier was supposedly sitting by his fire in his home of Avignon in 1782, watching embers rise off the flames. He theorized that this rising smoke contained a special property, which he dubbed "levity," and wondered if it could be harnessed. (Another parallel theory described the rising property of smoke as "phlogiston," an element released only through burning. Both theories are now obsolete.) Montgolfier and his brother, Jacques-Étienne, both paper-makers, began experimenting with capturing this heated gas within paper and cotton sacks, and the brothers are credited with inventing the hot-air balloon.

In a demonstration at Versailles before King Louis XVI and Queen Marie Antoinette on September 19, 1783, the brothers tested their invention by sending a sheep, a duck, and a chicken into the air. The flight lasted about eight minutes, covered a distance of nearly three kilometres, and reached an altitude of 500 metres. When the animals returned to the ground unscathed, the next obvious step was to send up a human.

Based on an extract from Andrew Bisharat, for *National Geographic*, published January 31, 2015

SCIENCE COMMUNICATION

The extract consists of an article from *National Geographic* newsfeed.

1 (a) Comment on the style of the writing, in particular the length of the article and level of scientific vocabulary used.

    (b) Give two reasons why the level of physics detail about the theories behind balloon flight is relatively low.

2 Explain the meaning of this quote from the text: 'Both theories are now obsolete'.

INTERPRETATION NOTE

Consider who is the intended audience. What do they most want to learn about extreme ballooning? This extract is only a small part of a much longer article online.

PHYSICS IN DETAIL

Now we will look at the physics in detail. Some of these questions will link to topics elsewhere in this book, so you may need to combine concepts from different areas of physics to work out the answers.

3 Explain why a detailed understanding of the gas laws governing the flight of the balloon is likely to be less important to these pilots than their knowledge of weather systems.

4 (a) Draw a free-body force diagram of the balloon at the moment of release at the ground.

    (b) Explain how the forces in your diagram in (a) have magnitudes that mean there is a resultant upward force.

5 This type of balloon has a fixed volume.

    (a) With reference to any relevant gas laws, explain why the pressure of the air outside the balloon falls as the balloon rises.

    (b) (i) This balloon, filled with helium, lifts off with a volume of $6\,200\,\text{m}^3$ at ground level ($20\,°\text{C}$ and a pressure of $101\,000\,\text{Pa}$) and rises to a maximum altitude of $9\,100\,\text{m}$ ($-44\,°\text{C}$ and a pressure of $30\,100\,\text{Pa}$). If no helium gas escapes, calculate the volume of the gas in the balloon at this new altitude.

        (ii) In fact, the maximum volume of the balloon at the altitude of $9\,100\,\text{m}$ is $9\,900\,\text{m}^3$. Explain why it does not expand to the volume you calculated in (b)(i).

    (c) Compare the average kinetic energy of air molecules at $9\,100\,\text{m}$ altitude with those at ground level.

6 As the air in a hot air balloon is heated, the balloon rises. Why did the smoke in Montgolfier's fireplace rise up the chimney?

PHYSICS TIP

Consider the ideal gas equation. Which quantities in the equation will remain constant during the flight?

ACTIVITY

Write a proposal for the extracurricular budget holder in your school outlining plans to set up a High Altitude Balloon Club in the school. The proposal should explain the basic idea of the formation of the club, and include an experimental purpose for your launch. You should include details of the timescale required for developing the equipment, a price list of the apparatus required, and information about how you will obtain necessary national and local permissions.

THINKING BIGGER TIP

Your proposal should encourage the budget holder to pay for the club and at least one balloon launch. Explain to the budget holder what benefits this project could bring for the school, for example, good publicity and an activity for as many students as possible. Note that a strong proposal will probably show evidence of minimised and well-controlled costs.

[Note: In questions marked with an asterisk (*), marks will be awarded for your ability to structure your answer logically, showing how the points that you make are related or follow on from each other.]

1 If a fixed mass of gas in a weather balloon moves to a new situation where the temperature is halved, and the pressure is also halved, what would happen to the volume of the balloon?

   A It would stay the same.

   B It would become a quarter of its original value.

   C It would halve.

   D It would double. [1]

   (Total for Question 1 = 1 mark)

2 How much energy is needed to melt 100 g of lead at 600 K?

   Specific latent heat of fusion of lead = 23 kJ kg^{-1}

   Melting point of lead = 327 °C

   A 2.3 J

   B 23 J

   C 2300 kJ

   D 2300 J [1]

   (Total for Question 2 = 1 mark)

3 The molar mass of oxygen molecules is 16 g. Assuming oxygen behaves as an ideal gas, what is the root-mean-square speed of oxygen molecules in a sample kept at 15 °C?

   A 21 m s^{-1}

   B 150 m s^{-1}

   C 670 m s^{-1}

   D 440 000 m s^{-1} [1]

   (Total for Question 3 = 1 mark)

4 (a) A typical aerosol can is able to withstand pressures up to 12 atmospheres before exploding. A 3.0×10^{-4} m^3 aerosol contains 3.0×10^{22} molecules of gas as a propellant. Show that the pressure would reach 12 atmospheres at a temperature of about 900 K.

   1 atmosphere = 1.0×10^5 Pa [2]

   *(b) Some aerosol cans contain a liquid propellant. The propellant exists inside the can as a liquid and a vapour. Explain what happens when such an aerosol can is heated to about 900 K. [3]

   (Total for Question 4 = 5 marks)

5 When your diaphragm contracts, the pressure in the chest cavity is lowered below atmospheric pressure and air is forced into your lungs.

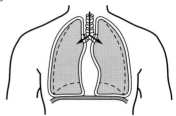

   (a) The diaphragm contracts and the lung capacity increases by 20%. State **two** assumptions you would need to make to calculate the new pressure in the lungs if the initial pressure is known. [2]

   (b) (i) The volume of air inhaled in a typical breath is 2.5×10^{-4} m^3 and an adult takes about 25 breaths per minute. Show that the mass of air taken into the lungs each second is about 1×10^{-4} kg.

   Density of air = 1.2 kg m^{-3} [2]

   (ii) Body temperature is 37.6 °C and the temperature outside the body is 20.0 °C. Calculate the rate at which energy is used to warm air up to body temperature.

   Specific heat capacity of air = 1000 J kg^{-1} K^{-1} [2]

   (Total for Question 5 = 6 marks)

6 A student uses the apparatus shown to investigate the relationship between pressure and volume of a gas.

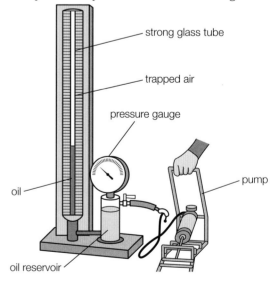

strong glass tube

trapped air

pressure gauge

pump

oil

oil reservoir

Air is trapped in a glass tube of uniform cross-sectional area. As the pressure of the trapped air is increased, the length of trapped air decreases. The student collects data and plots the following graph.

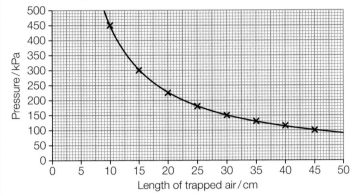

(a) State the variables that should be controlled in this investigation. [2]

(b) Theory suggests that, for the air trapped in the tube, the pressure p is inversely proportional to the volume V. Use the graph to show that this relationship is correct. State an assumption that you are making. [4]

(c) On the day that the investigation was carried out, the temperature in the laboratory was 20 °C.
Calculate the number of air molecules trapped in the tube.
cross-sectional area of tube = $7.5 \times 10^{-5} \, m^2$ [3]

(d) State how the graph would change if
  (i) the air molecules in the tube were replaced by the same number of molecules of hydrogen gas. [1]
  (ii) the temperature of the laboratory was much higher. [2]
  (Total for Question 6 = 12 marks)

7 A student wants to determine the specific heat capacity of aluminium. He heats a block of aluminium by supplying electrical energy to a heater that is inserted into the block as shown.

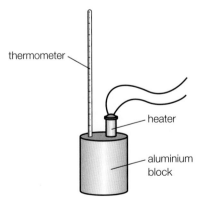

(a) Draw the electrical circuit he should use. [1]

(b) He writes the following plan:

Measure the temperature of the block at the start

Turn on the electric current

Determine the energy flowing for a certain time

Turn off the current and measure the temperature

This will give me the specific heat capacity if I divide the total energy by the mass and temperature rise

Write an improved plan that includes details of the method to be used and any precautions needed to produce an accurate value for the specific heat capacity of aluminium. [6]
  (Total for Question 7 = 7 marks)

8 The picture shows an inflatable globe. This is a flexible plastic sphere on which a map of the world is printed. It is inflated by blowing into it like a balloon.

When fully inflated the globe has a volume of $6.55 \times 10^{-2} \, m^3$. At a temperature of 22 °C the pressure exerted by the air in the globe is $1.05 \times 10^5 \, Pa$.

(a) On average there are 1.25×10^{22} molecules in each breath of air that we take. Show that the number of breaths needed to fully inflate the globe is about 140. [3]

(b) The fully inflated globe is left outside and its temperature rises from 22 °C to 30 °C. The volume of the globe remains constant.

  Calculate the new pressure exerted by the air in the globe. [2]

(c) Including ideas of momentum, explain why the pressure exerted by the air in the globe increases. [4]
  (Total for Question 8 = 9 marks)

TOPIC 9 NUCLEAR DECAY

CHAPTER 9A RADIOACTIVITY

Radiation is scary. At least, that is what most people believe. When the USA dropped nuclear bombs on Japan in 1945, the Japanese suffered so many casualties that they surrendered within days. However, there is radioactivity naturally all around us. Having a picnic on a rock in a national park could easily expose you to more radioactivity than undertaking experiments in school using radioactive sources. Coal-fired power stations release more radioactivity to the environment, but nuclear power stations emit more radioactivity from their fuel. It is important to understand the evidence which relates to hazards in all areas of life. However, nuclear physics is poorly understood by the public and its hazards are often overstated.

In this chapter, we will learn what radioactivity is, how it can be dangerous, and how these hazards can be reduced. We will also see how the energy stored in the nuclei of atoms can be measured, where it comes from, and how it can be used as a power source.

MATHS SKILLS FOR THIS CHAPTER

- **Changing the subject of an equation** (*e.g. finding the mass in $E = mc^2$*)
- **Use of exponential and logarithmic functions** (*e.g. solving for unknowns in the radioactive decay equation*)
- **Use of the principles of calculus** (*e.g. solving the rate of decay differential equation*)
- **Use of logarithmic graph plots** (*e.g. finding the half-life from experimental data from a linear log plot*)
- **Calculating surface areas** (*e.g. verifying the inverse square law for gamma ray intensity*)

What prior knowledge do I need?

Topics 1B (Book 1: IAS) and 7B

- The conservation of energy

Topic 3D (Book 1: IAS)

- Ionisation
- Photon emission from excited electrons

Topic 7A

- The structure of an atom
- Rutherford's alpha particle scattering experiment

Topic 7C

- Particle energies

What will I study in this chapter?

- The different types of radioactive emissions
- Mass defect and binding energy
- The importance of binding energy per nucleon
- Nuclear fission and nuclear fusion
- Nuclear reactions
- The random nature of radioactive decay, and radioactive half-life

What will I study later?

Topic 11A

- How gravitational forces follow an inverse square relationship

Topic 11B

- How nuclear fusion reactions power stars
- Gamma ray bursts from supernova explosions
- The production of the heaviest elements in supernovae

LEARNING OBJECTIVES

■ Describe what is meant by *background radiation* and explain how to take appropriate account of it.

■ Explain the relationship between the structure of alpha, beta and gamma radiations, and their penetrating and ionising abilities.

■ Write and interpret nuclear equations.

BACKGROUND RADIATION

Human beings can survive small doses of nuclear radiation. This has been important in our evolution, as the natural environment has low levels of radiation from natural sources. This is called **background radiation**. The level varies from place to place around the world. For example, in the UK it averages to less than one radioactive particle every two seconds in any given place (**fig A**). If we measure background radiation using a Geiger–Müller (G–M) tube, the number of counts per second usually ranges from 0.2–0.5 depending on exact location. Radiation levels are often reported in counts per second, and this unit is called the becquerel (Bq). Henri Becquerel was the French physicist who discovered spontaneous radiation in 1896.

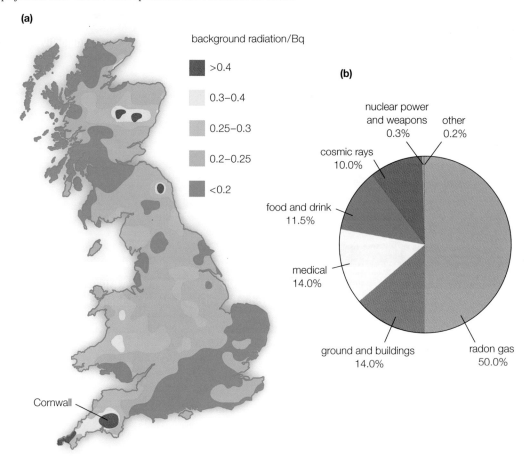

▲ **fig A** The environment continually exposes us to low levels of nuclear radiation. (a) Levels of background radiation in Great Britain (measured in Bq); (b) Sources of background radiation.

The exact exposure to nuclear radiations that any individual will receive from their environment will depend on where they are and how long they are there. This is because different environmental factors contribute to the local level of background radiation. On Earth, approximately half of the background radiation is from naturally radioactive gases which are in the atmosphere, particularly radon. Radon is produced in the decay of uranium ore present in certain rocks (especially granite) and so is found more in certain parts of the world than others. In Cornwall in the UK, some houses are fitted with radon detectors and special ventilation systems to remove excess radon gas from the household atmosphere.

DID YOU KNOW?

The most naturally radioactive place on Earth is Ramsar, in Iran, where the dosage can be more than 200 times the global average.

PRACTICAL SKILLS

Investigating background radiation

It is easy to determine the average background radiation in your area using a Geiger–Müller tube and counter. As radioactive decay is a random and spontaneous process, the activity in your lab must be measured over a long period of time (30 minutes or more) and then an average calculated. Otherwise, you may find that the measurement is, by chance, particularly high or particularly low and so does not truly indicate the average over time. For example, a good approximation to the average will be achieved if the G–M tube and counter are set to counting for two hours. The final count is then divided by 7200 (seconds), which is long compared with the average count of about 0.5 Bq. Measurements of radioactivity that have had the background radiation deducted, so that they only represent activity by a radioactive source under test, are known as *corrected counts*.

WORKED EXAMPLE 1

In a school experiment, Ricardo measured the background radiation using a G–M tube and counter for half an hour. The final reading was 747 counts. What is the average background radiation?

$$\text{time} = 30 \times 60 = 1800\,\text{s}$$
$$\text{activity} = \frac{747}{1800} = 0.415\,\text{Bq}$$

The background count will skew the results of investigations into nuclear radiations. Whenever such an investigation is undertaken, the background radiation must also be measured separately and then deducted from each count measured in the main part of the investigation.

TYPES OF NUCLEAR RADIATION

Many nuclei are slightly unstable and there is a small probability that, each second, they may **decay**. This means that a nucleon may change from one type to another, or the composition or energy state of the nucleus as a whole may change. When a nuclear decay occurs, the radiation particle emitted will leave the nucleus carrying kinetic energy. As the particle travels, it will ionise particles in its path, losing some of that kinetic energy at each ionisation. When all the kinetic energy is transferred, the radiation particle stops and is absorbed by the substance that it is in at that moment.

The three main types of nuclear radiation are called **alpha (α), beta (β)** and **gamma (γ)** radiation. Each one comes about through a different process within the nucleus, each one is composed of different particles, and each one has different properties.

PRACTICAL SKILLS

Investigating radiation penetration

You can investigate the penetrating power of alpha, beta and gamma radiation using a Geiger–Müller tube to detect them. Between the source and the G–M tube, place absorber sheets that progressively increase in density, and measure the average count rate. This investigation is often simulated using computer software. This removes all risk of exposure to radiological hazards (see below).

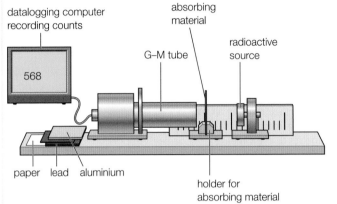

▲ **fig B** Investigating nuclear radiations.

Safety Note: Handle the radioactive sources with a long pair of tongs and do not point the open end of the source at any part of the body. Return each source to its secure box after use.

ALPHA DECAY

Alpha particles are composed of two protons and two neutrons, the same as a helium nucleus. This is a relatively large particle with a significant positive charge ($+2e$), so it is highly ionising. An alpha particle moving through air typically causes 10 000 ionisations per millimetre. As it ionises so much, it quickly loses its kinetic energy and is easily absorbed. A few centimetres travel in air is enough to absorb an alpha particle, and they are also blocked by paper or skin.

Example alpha decay equation: $^{241}_{95}\text{Am} \rightarrow {}^{237}_{93}\text{Np} + {}^{4}_{2}\alpha$

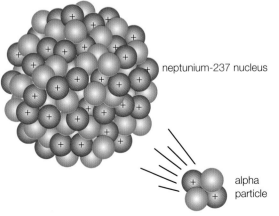

▲ **fig C** Alpha particle emission.

BETA DECAY

A beta particle is an electron emitted at high speed from the nucleus when a neutron decays into a proton. With its single negative charge and much smaller size, the beta particle is much less ionising than an alpha particle, and thus can penetrate much further. Several metres of air, or a thin sheet of aluminium, are needed to absorb beta particles.

Example beta decay equation:

$$^{14}_{6}\text{C} \rightarrow ^{14}_{7}\text{N} + ^{0}_{-1}\beta + ^{0}_{0}\overline{\upsilon}$$

A neutron changes into a proton and an electron.

unstable carbon-14 nucleus

beta particle

nitrogen nucleus

▲ **fig D** Beta-minus particle emission.

GAMMA DECAY

Gamma rays are high energy, high frequency, electromagnetic radiation. These photons have no charge and no mass and so will rarely interact with particles in their path, which means they are the least ionising nuclear radiation. They are never completely absorbed, although their energy can be significantly reduced by several centimetres of lead or several metres of concrete. If the energy of gamma rays is reduced to a safe level, we often describe them as being absorbed.

Example gamma decay equation:

$$^{60}_{28}\text{Ni-m} \rightarrow ^{60}_{28}\text{Ni} + ^{0}_{0}\gamma$$

gamma ray

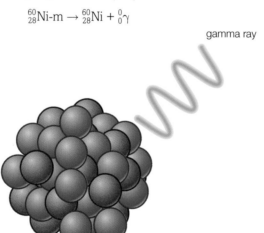

▲ **fig E** Gamma ray emission. The image shows only the gamma emission from an excited nickel nucleus, Ni-m a moment after a beta particle was emitted from a cobalt-60 nucleus.

Gamma ray emission does not change the structure of a nucleus emitting it, but carries away energy, so the nucleus must drop energy level in order to emit the gamma ray. This most commonly happens immediately after alpha or beta decay, which leaves the nucleus in an excited state. The emission of the gamma ray photon and associated de-excitation of the nucleus is exactly the same as we saw in **Book 1, Section 3D.4** for the production of visible light, but involves nuclear energy states and not electrons. The difference is that the amount of energy involved is many times higher, so this produces much higher frequency gamma radiation.

PRACTICAL SKILLS

Investigating lead absorption of gamma radiation

Using the same experimental set-up as shown in **fig B**, we can find out how much lead is needed to reduce the intensity of gamma radiation by a particular amount. Using a radioactive source that emits alpha and gamma only, we can select to study the gamma radiation by placing a sheet of paper in front of the source to absorb all the alpha particles. This will have no measurable effect on the gamma emissions.

Keeping the distance from the source to the G–M tube constant, we can place different thicknesses of lead plate in the gap between them. For each thickness, record the count over a fixed time period and determine the corrected count rate. There is a mathematical relationship between the corrected count rate (which represents intensity of the radiation, I) and the thickness, x, of lead. Work out what graph to draw to prove that $I \propto e^{-\mu x}$.

Safety Note: Handle the gamma source with a long pair of tongs and do not point the open end at any part of the body. Be careful not to put fingers and hands in front of the source when changing the lead blocks.

DANGERS FROM NUCLEAR RADIATIONS

Ionising radiations can interact with the particles that human cells consist of. There may be so much ionisation that it causes the cells to die. Where there is less ionisation, the molecules of DNA in the cells may change slightly. These DNA mutations can cause cells to have an increased tendency to become cancerous. As the different types of nuclear radiation ionise to different extents, the hazard to humans is different for each type. The hazard levels are given in **table A**.

TYPE OF RADIATION	INSIDE BODY	OUTSIDE BODY
alpha, α	Highly ionising – very dangerous: radiation poisoning and cancer possible.	Absorbed by surface layer of dead skin cells – no danger.
beta, β	Moderate ionisation and danger: exposure should be minimised.	Moderate ionisation and danger: close exposure should be minimised.
gamma, γ	Minimal ionisation – cancer danger from long-term exposure.	Minimal ionisation – cancer danger from long-term exposure.

table A Hazards from nuclear radiations. Not considered here is the fact that different tissue types can be more or less affected by these dangers.

CHECKPOINT

SKILLS ▶ INNOVATION

1. ▶ Create a table summarising everything you know about alpha, beta and gamma radiations.

2. In preparation for an experiment using a radioactive sample, Xian measured the radioactivity in the laboratory without the sample present. In one hour, the Geiger–Müller tube measured 1645 counts. What was the background count in becquerel?

3. Why is it theoretically safe to hold a sample which emits only alpha radiation? Why should you still never do so?

4. Why might the UK government list 'living in Cornwall' as one of the most (radiologically) hazardous activities that the public can undertake?

5. ▶ It is thought that some soil could be contaminated with a radioisotope. You have a sample of this soil. Describe an experiment to determine what types of radiation are emitted.

9A 2 RATE OF RADIOACTIVE DECAY

LEARNING OBJECTIVES

- Describe the spontaneous and random nature of radioactive decay.
- Determine half-lives of isotopes graphically.
- Use the equations for rate of radioactive decay, and their logarithmic equivalents.

PROBABILITY AND DECAY

Radioactive decay is a spontaneous and random process and so any radioactive nucleus may decay at any moment. For each second that it exists, there is a particular probability that the nucleus will decay. This probability is called the **decay constant**, λ. However, just like guessing which number will come up next in a lottery, it is not possible to predict when any given nucleus will decay. The chance that a particular nucleus will decay is not affected by factors outside the nucleus, such as temperature or pressure, or by the behaviour of neighbouring nuclei – each nucleus decays entirely independently.

If we have a large sample of the nuclei, the probability of decay will determine the fraction of these nuclei that will decay each second. Naturally, if the sample is larger, then the number that decay in a second will be greater. So the number decaying per second, called the **activity**, A $\left(\text{or } \dfrac{dN}{dt}\right)$, is proportional to the number of nuclei in the sample, N. Mathematically, this is expressed as:

$$A = -\lambda N$$

$$\frac{dN}{dt} = -\lambda N$$

The minus sign in this formula occurs because the number of nuclei in the sample, N, decreases with time. In practice we ignore the negative sign when using the formula.

WORKED EXAMPLE 1

The decay constant for carbon-14 is $\lambda = 3.84 \times 10^{-12}\,\text{s}^{-1}$. What is the activity of a sample of 100 billion atoms of carbon-14?

$$\frac{dN}{dt} = -\lambda N$$

$$\frac{dN}{dt} = -(3.84 \times 10^{-12}) \times (100 \times 10^9) = 0.384\,\text{Bq}$$

The formula for the rate of decay of nuclei in a sample is a differential equation. We have previously met this type of equation when studying the discharge of a capacitor. The equation $\dfrac{dN}{dt} = -\lambda N$ can be solved to give a formula for the number of nuclei remaining in a sample, N, after a certain time, t

$$N = N_0\,e^{-\lambda t}$$

where N_0 is the initial number of nuclei within a sample.

WORKED EXAMPLE 2

If our sample of 100 billion carbon-14 atoms were left for 300 years, how many carbon-14 atoms would remain?

$$\text{time, } t = 300 \times 365 \times 24 \times 60 \times 60 = 9.46 \times 10^9\,\text{s}$$

$$N = N_0\,e^{-\lambda t}$$

$$\lambda t = (3.84 \times 10^{-12})(9.46 \times 10^9) = 0.0363$$

$$N = (100 \times 10^9) \times e^{-0.0363}$$

$$\therefore \quad N = 9.64 \times 10^{10}\,\text{atoms}$$

LEARNING TIP

The exponential mathematics that govern radioactive decay is identical in structure to those for the discharge of capacitors, which we saw in **Section 6B.3**.

HALF-LIFE

As we have seen, the activity of a radioactive sample decreases over time as the radioactive nuclei decay, leaving fewer radioactive nuclei available to decay. While the activity of a sample depends on the number of nuclei present, the rate at which the activity decreases depends only on the particular isotope. A measure of this rate of decrease of activity is the **half-life**, $t_{\frac{1}{2}}$. This is the time taken for half of the nuclei of that nuclide within a sample to decay.

Mathematically, the half-life can be found by putting $N = \frac{1}{2}N_0$ into the decay equation:

$$N = N_0\,e^{-\lambda t}$$

$$\frac{N_0}{2} = N_0\,e^{-\lambda t_{\frac{1}{2}}}$$

$$\frac{1}{2} = e^{-\lambda t_{\frac{1}{2}}}$$

$$\therefore \quad \ln\left(\tfrac{1}{2}\right) = -\lambda t_{\frac{1}{2}}$$

$$-\ln 2 = -\lambda t_{\frac{1}{2}}$$

$$t_{\frac{1}{2}} = \frac{\ln 2}{\lambda}$$

Rearranging this also gives us:

$$\lambda = \frac{\ln 2}{t_{\frac{1}{2}}}$$

WORKED EXAMPLE 3

What is the half-life of carbon-14?

$$\lambda = 3.84 \times 10^{-12}\,\text{s}^{-1}$$

$$t_{\frac{1}{2}} = \frac{\ln 2}{\lambda}$$

$$t_{\frac{1}{2}} = \frac{\ln 2}{3.84 \times 10^{-12}}$$

$$t_{\frac{1}{2}} = 1.81 \times 10^{11}\,\text{s} = 5720\,\text{years}$$

Investigating radioactive decay rates

You may have the equipment to measure the half-life for a radioactive sample, as shown in **fig A**. If you do not, a simulation in which dice represent the radioactive nuclei can demonstrate the exponential decay of a sample (**fig B**).

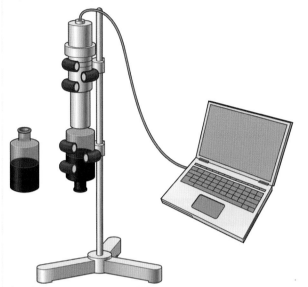

▲ **fig A** Measuring the half-life of protactinium-234m.

> **!**
> Safety Note: Before use, check that the bottle is not leaking and never remove the cap. Put a tray under the bottle in case it does leak when in use. Store it in a larger plastic container with a lid.

▲ **fig B** Measuring the half-life of dice.

HALF-LIFE GRAPHS

An experiment to determine the half-life of a substance will usually measure its activity over time. Activity is proportional to the number of nuclei present, so when the activity is plotted against time, the shape of the curve is exponential decay (**fig C**). The activity, A, follows the equation:

$$A = A_0 e^{-\lambda t}$$

We can use the graph of activity against time to determine the half-life of the substance by finding the time taken for the activity to halve.

To find the half-life from such a graph, find a useful start point on the curve. In the example shown in **fig C**, this could be the start point where the activity is 800 Bq. As the half-life is defined as the time taken for the activity to fall to half of its original value, we use the graph to find the time taken for the count to drop to 400 Bq, which is 70 seconds. Doing this a second time, from a count rate of 400 Bq to a count rate of 200 Bq, gives a time of 80 seconds. Notice that the time interval is not identical each time. This is due to the random nature of radioactive decay, plus experimental and graphing errors. The best-fit curve will be a matter of the drawer's judgement. Thus, to get the best answer for the half-life, we must undertake the analysis on the graph several times in different parts of the graph and average the results. From the two half-lives shown in **fig C**, the average half-life for protactinium-234m would be 75 seconds.

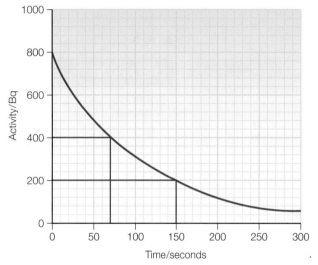

▲ **fig C** Determining half-life graphically.

LOGARITHMIC HALF-LIFE ANALYSIS

As we have seen in **fig C**, there are many graphical uncertainties which could mean that our conclusion of the half-life for a sample under test is unreliable. To improve on the curving graph, we can analyse the experimental data to generate a straight-line graph so our best-fit line will be less uncertain. Using either of our decay equations, taking the logarithm of the equation will give us a new equation for the data that is in the form $y = mx + c$.

Considering data from measuring the activity of a sample over time:

$$A = A_0 e^{-\lambda t}$$

$$\ln A = \ln A_0 - \lambda t$$

$$\therefore \quad \ln A = -\lambda t + \ln A_0$$

So a plot of $\ln A$ on the y-axis against t on the x-axis (**fig D**) will be a straight line with a negative gradient equal in magnitude to the decay constant, λ. The y-intercept will be the natural logarithm value of the initial activity.

Alternatively, if the experimental data is measuring number of nuclei (often actually measured as mass):

$$N = N_0 e^{-\lambda t}$$

$$\ln N = \ln N_0 - \lambda t$$

$$\therefore \quad \ln N = -\lambda t + \ln N_0$$

Here, a plot of $\ln N$ on the y-axis against t on the x-axis will also be a straight line with a negative gradient, which is equal in magnitude to the decay constant, λ. The y-intercept in this case will be the natural logarithm value of the initial number of nuclei.

In either case, the gradient gives the decay constant, from which we can find the half-life from:

$$t_{\frac{1}{2}} = \frac{\ln 2}{\lambda}$$

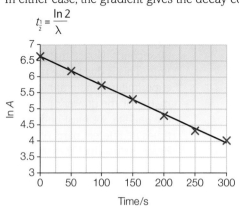

▲ **fig D** Logarithmic graphical analysis of protactinium-234m decay data.

CHECKPOINT

SKILLS **ANALYSIS**

1. ▶ Calculate the decay constants for the following isotopes that are commonly used in school laboratories:

   (a) radium-226: half-life = 1602 years

   (b) strontium-90: half-life = 28.8 years

   (c) cobalt-60: half-life = 5.3 years.

2. If the cellar in a house in Nova Scotia, Canada contained 6.5 billion atoms of radon-222 gas, with a decay constant $\lambda = 2.10 \times 10^{-6}\,\text{s}^{-1}$, how many radon gas atoms would there be one day later?

3. ▶ Technetium-99m is a gamma emitter which is often used as a medical tracer to monitor lymph node activity. Use the graph of experimental results shown in **fig E** to work out the half-life of technetium-99m.

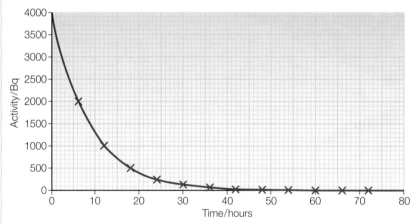

▲ **fig E** Radioactive decay of technetium-99m.

4. Explain why a radioactive isotope with a very small decay constant will have a long half-life.

5. ▶ Find the half-life of protactinium from the logarithmic graph in **fig D**.

SUBJECT VOCABULARY

decay constant the probability, per second, that a given nucleus will decay

activity the number of radioactive decays in unit time

half-life the time taken for half of the nuclei within a sample to decay. Alternatively, the time taken for the activity of a sample of a radioactive nuclide to reduce to half its initial value

ENERGY–MASS EQUIVALENCE

A nucleon has a mass which is approximately equal to 1 unified atomic mass unit, u ($= 1.66 \times 10^{-27}$ kg). The exact masses of a proton, a neutron and an electron are given in **table A**.

PARTICLE	MASS/ATOMIC MASS UNITS/U	MASS/SI UNITS/KG
proton	1.007 276	$1.672\,623 \times 10^{-27}$
neutron	1.008 665	$1.674\,929 \times 10^{-27}$
electron	0.000 548 58	$9.109\,390 \times 10^{-31}$

table A Masses of sub-atomic particles.

We might expect that if we know the constituent parts of any nucleus, we can calculate its mass by finding the total mass of its nucleons. However, in practice we find that the actual, measured mass of a nucleus is always *less* than the total mass of its constituent nucleons. This difference is called the **mass deficit, Δm**, or sometimes the **mass defect, Δm**.

WORKED EXAMPLE 1

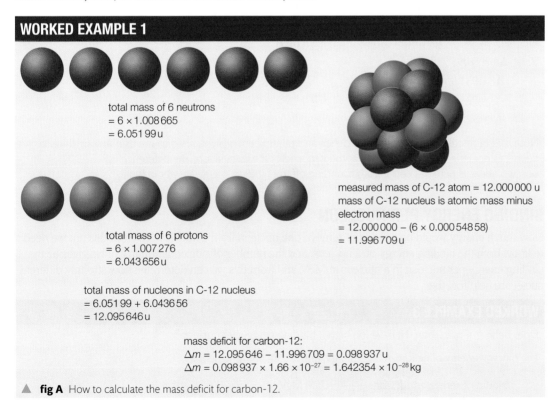

total mass of 6 neutrons
$= 6 \times 1.008\,665$
$= 6.051\,99\,\text{u}$

total mass of 6 protons
$= 6 \times 1.007\,276$
$= 6.043\,656\,\text{u}$

measured mass of C-12 atom = 12.000 000 u
mass of C-12 nucleus is atomic mass minus electron mass
$= 12.000\,000 - (6 \times 0.000\,548\,58)$
$= 11.996\,709\,\text{u}$

total mass of nucleons in C-12 nucleus
$= 6.051\,99 + 6.043\,656$
$= 12.095\,646\,\text{u}$

mass deficit for carbon-12:
$\Delta m = 12.095\,646 - 11.996\,709 = 0.098\,937\,\text{u}$
$\Delta m = 0.098\,937 \times 1.66 \times 10^{-27} = 1.642\,354 \times 10^{-28}\,\text{kg}$

▲ **fig A** How to calculate the mass deficit for carbon-12.

NUCLEAR BINDING ENERGY

The mass deficit comes about because a small amount of the mass of the nucleons is converted into the energy needed to hold the nucleus together. This is called **binding energy, ΔE**. It is calculated using Einstein's mass–energy relationship:

$$\Delta E = c^2 \Delta m$$

where c is the speed of light.

There are two common systems of units for calculating binding energy. If you have calculated the mass deficit in kilograms
(SI units) then using $c = 3.00 \times 10^8 \, \text{m s}^{-1}$ will give the binding energy in joules. Alternatively, if you have calculated the mass deficit in atomic mass units, then you convert this into binding energy in mega-electronvolts (MeV) using:

$$1 \, \text{u} = 931.5 \, \text{MeV}$$

WORKED EXAMPLE 2

Calculate the binding energy for a carbon-12 nucleus in both joules and electronvolts.

Working in kilograms, from **fig A**:

$$\Delta m = 1.642\,354 \times 10^{-28} \, \text{kg}$$

$$\Delta E = c^2 \Delta m$$
$$= (3 \times 10^8)^2 \times (1.642\,354 \times 10^{-28})$$
$$= 1.478 \times 10^{-11} \, \text{J}$$

Converting to electronvolts:

$$\Delta E = \frac{1.478 \times 10^{-11}}{1.6 \times 10^{-19}}$$
$$= 9.226 \times 10^7 \, \text{eV}$$
$$= 92.3 \, \text{MeV}$$

Alternatively, working in atomic mass units, from **fig A**:

$$\Delta m = 0.098\,937 \, \text{u}$$

$$\Delta E = 931.5 \times \Delta m$$
$$= 931.5 \times 0.098\,937$$
$$= 92.2 \, \text{MeV}$$

Note that in calculations of both mass deficit and binding energy, you need to use as many significant figures as possible, only rounding off at the very end of the calculation. The difference in our two answers here has come from rounding off the binding energy answer in joules.

BINDING ENERGY PER NUCLEON

How much energy would be needed to remove one nucleon from a nucleus? To work this out we need to know both the binding energy of a nucleus, and the number of nucleons within it. This gives us the binding energy per nucleon in a nucleus in MeV, and from this we can determine how strongly different nuclei are held together.

WORKED EXAMPLE 3

Does helium-4 or carbon-12 have the higher binding energy per nucleon?

The binding energy of helium-4 is 28.3 MeV, and it contains 4 nucleons:

$$\text{binding energy per nucleon} = \frac{28.3}{4} = 7.08 \, \text{MeV}$$

The binding energy for carbon-12 is 92.2 MeV, and it contains 12 nucleons:

$$\text{binding energy per nucleon} = \frac{92.2}{12} = 7.68 \, \text{MeV}$$

So, carbon-12 nuclei are more tightly bound together than those of helium-4.

Drawing a graph of binding energy per nucleon against mass number for the nuclei gives us a useful way of comparing how tightly different nuclides are bound together.

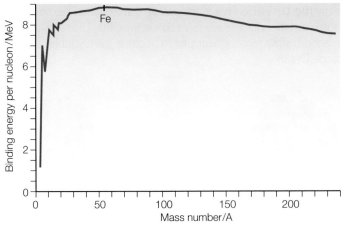

▲ **fig B** Graph of binding energy per nucleon against mass number, A.

Looking at **fig B**, you will see that the isotope with the highest binding energy per nucleon is iron-56 at 8.8 MeV per nucleon. Any nuclear reaction which increases the binding energy per nucleon will give out energy.

The graph shows us that small nuclides can combine together to make larger nuclei (up to Fe-56) with a greater binding energy per nucleon. This process is called nuclear **fusion**. Similarly, larger nuclei can break up into smaller pieces which have a greater binding energy per nucleon than the original nucleus. Reactions like this are called nuclear **fission**. Both of these types of nuclear reaction will give out energy, and could be used as power sources.

NUCLEAR FUSION

If we take some light nuclei and force them to join together, the mass of the new heavier nucleus will be less than the mass of the constituent parts, as some mass is converted into energy. However, not all of this energy is used as binding energy for the new larger nucleus, so energy will be released from this reaction. The binding energy per nucleon afterwards is higher than at the start. This is the process of nuclear fusion and is what provides the energy to make stars shine (**fig C**).

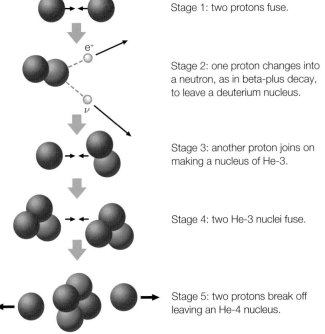

Stage 1: two protons fuse.

Stage 2: one proton changes into a neutron, as in beta-plus decay, to leave a deuterium nucleus.

Stage 3: another proton joins on making a nucleus of He-3.

Stage 4: two He-3 nuclei fuse.

Stage 5: two protons break off leaving an He-4 nucleus.

▲ **fig C** The proton–proton nuclear fusion reaction, typical in small cool stars such as our Sun, where the core temperature is about 15×10^6 K.

WORKED EXAMPLE 4

The proton–proton chain nuclear fusion reaction effectively takes four protons and converts them into a helium-4 nucleus and two positrons (which have the same mass as electrons). Calculate the energy released in this reaction.

The measured *atomic* mass of ^{4}He is 4.002 602 u.

Mass of 4 protons = 4 × 1.007 276 u
= 4.029 104 u

Mass of ^{4}He *nucleus* + 2 positrons
= (4.002 602 − 2 × 0.000 548 58) u + (2 × 0.000 548 58) u
= 4.002 602 u

Mass deficit = 4.029 104 u − 4.002 602 u
= 0.0265 u
= 24.7 MeV

These reactions which occur in enormous numbers provide the energy which causes the Sun to emit heat and light in all directions in great quantities.

Converting hydrogen into helium with the release of energy could be an excellent way of supplying the planet's energy needs. The seas are full of hydrogen in water molecules, and the helium produced would be an inert gas which could simply be allowed to go up into the upper atmosphere. However, scientists have not yet successfully maintained a controlled nuclear fusion reaction. The problem lies in forcing two positively charged, mutually repelling, protons to fuse together. The kinetic energy which they need to collide forcefully enough to overcome this electrostatic repulsion requires temperatures of many million kelvin. Moreover, to ensure enough colliding protons for the reaction to be sustained requires a very high density of them. Comparing the energy output with nuclear fission below, 235 grammes of hydrogen undergoing nuclear fusion would produce an energy output of 1.40×10^{14} J.

FUSION REACTIONS

Nuclear fusion as an energy source for us to generate electricity is currently only at a research stage. Fusion would have the advantage that its fuel source is abundant, and no radioactive waste would be produced. It is difficult to achieve though, because the high temperature and density needed make confining the 'burning plasma' difficult. This confinement would need very strong magnetic fields. The ITER project (International Thermonuclear Experimental Reactor) is a research experiment for nuclear fusion to produce electricity. This is an international project, ITER will be sited in France, and the EU and six other large nations (India, China, Russia, USA, Korea and Japan) will also contribute towards the 15 billion Euros construction cost. An estimated 10 years of construction began in 2010. By 2025, the aim is to generate temperatures of 150 million kelvin in a plasma volume of 840 cubic metres.

NUCLEAR FISSION

We have seen that nuclear fusion is not yet an option for electricity generation, as we have not been able to create the high densities and temperatures needed to sustain a fusion reaction. However,

another process which releases binding energy from nuclei is called **nuclear fission**. In this process a large nucleus breaks up into two smaller nuclei, with the release of some neutrons and energy (**fig D**). Fission reactions can be started when the nucleus absorbs another particle which makes it unstable. Uranium-235 can absorb slow-moving neutrons to become the unstable isotope U-236.

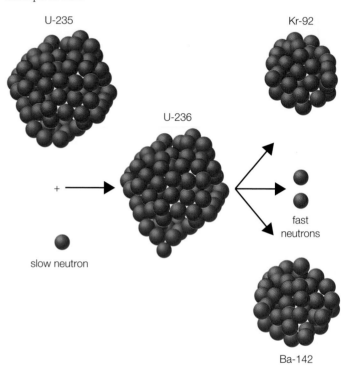

▲ **fig D** Nuclear fission of uranium-235.

WORKED EXAMPLE 5

Although there are many different possible products from nuclear fission reactions, the amount of energy released in each one is about the same at about 200 MeV.

Calculate the energy released in the following fission reaction:

$$^{235}U + n \rightarrow {}^{141}Ba + {}^{92}Kr + 3n + energy$$

(Data: mass of U-235 = 235.0439 u; mass of Ba-141 = 140.9144 u; mass of Kr-92 = 91.9262 u.)

Mass difference: Δm
= 235.0439 + 1.008 665 − 140.9144 − 91.9262 − 3(1.008 665)
= 0.1860 u

Energy released: E = 0.1860 × 931.5
= 173 MeV
= 2.77×10^{-11} J

This is the energy per fission. If one mole of these were to occur, using 235 g of U-235, then the total energy produced would be:

$E = 6.02 \times 10^{23} \times 2.77 \times 10^{-11}$
= 1.67×10^{13} J

However, U-235 is a small proportion (≈ 0.7%) of all the uranium found, and thus a larger amount of uranium fuel would be needed in order to provide enough U-235 atoms to produce this much energy.

What are the changes in binding energy per nucleon in the above reaction?

(Data: binding energy of U-235 = 1786 MeV; binding energy of Ba-141 = 1170 MeV; binding energy of Kr-92 = 782 MeV.)

Find the binding energies per nucleon:

U-235: $\dfrac{1786}{235}$ = 7.6 MeV

Ba-141: $\dfrac{1170}{141}$ = 8.3 MeV

Kr-92: $\dfrac{782}{92}$ = 8.5 MeV

Therefore the two fission products each have a higher binding energy per nucleon than the original uranium nuclide.

DID YOU KNOW?

NUCLEAR BOMBS

The neutrons in controlled nuclear fission reactions are slowed to speeds needed to sustain fission using a moderator. However, some nuclear fission reactions, such as those using plutonium, will progress by reacting with high speed neutrons, meaning that a moderator is not required. If such a reaction is allowed to run uncontrolled, it produces energy continuously, at an ever-increasing rate, until all the fuel is used up. This is the essential concept behind the design of nuclear bombs. A lump of plutonium-239 about the size of a tennis ball can completely react in less than a microsecond, releasing the energy equivalent of 20 kilotonnes of TNT, about 90×10^{12} J.

This type of bomb (fig E) was dropped on the city of Nagasaki in Japan during the Second World War. It killed 40 000 people immediately, and a similar number had died from after-effects by the end of 1945. The atomic bombings in Japan were so devastating that the country surrendered to the allies just six days after the Nagasaki bomb, and since the 1960s the Japanese government has resolved never to allow Japan to hold nuclear weapons.

▲ fig E A Fat Man plutonium fission bomb.

CHECKPOINT

SKILLS INTERPRETATION, ANALYSIS

1. ▶ Use the data in **table A** to work out how much energy could be produced if a proton, neutron and electron were each completely converted into energy.

2. The atomic mass unit was different between physics and chemistry until the system was unified in the 1960s. Previously, in Physics, it was based on the mass of an atom of oxygen-16, which is 15.994 915 u (now). Using the modern values, calculate the binding energy of a nucleus of oxygen-16.

SKILLS CRITICAL THINKING

3. ▶ Inside large mass stars, fusion of heavier elements than in the proton–proton cycle can occur. At about 10^8 K, helium-4 nuclei will fuse into carbon-12 nuclei according to the equation:

$$^4He^{2+} + {}^4He^{2+} + {}^4He^{2+} = {}^{12}C^{6+}$$

(a) Calculate the energy released in this reaction.
(b) How does the binding energy per nucleon change in the reaction?
(c) What is the average kinetic energy and rms speed of the helium nuclei in the star?

4. Complete the following nuclear fission reaction to determine X:

$$^{235}U + n \rightarrow {}^{X}Rb + {}^{139}Cs + 2n + energy$$

5. Calculate the amount of energy, in joules, generated from 2 kg of uranium fuel, if the U-235 represents 0.7% of the metal and every fission reaction produces 200 MeV.

6. Calculate the energy released in one fission in the following reaction:

$$^{235}U + n \rightarrow {}^{94}Zr + {}^{139}Te + 3n + energy$$

(Data: mass of Zr-94 = 93.9063 u; mass of Te-139 = 138.9347 u.)

7. Explain, in terms of binding energy per nucleon, why nuclear fusion and nuclear fission can release energy.

SUBJECT VOCABULARY

mass deficit, Δm the difference between the measured mass of a nucleus and the sum total of the masses of its constituent nucleons

mass defect, Δm an alternative phrase for 'mass deficit'

binding energy, ΔE the energy used to hold the nucleus together, converted from the mass deficit, following $E = mc^2$. So, it is also the energy needed to break a nucleus apart into its individual nucleons

fusion small nuclides combine together to make larger nuclei, releasing energy

fission larger nuclei are broken up into smaller nuclides, releasing energy

nuclear fission a large nucleus breaks up into two smaller nuclei, with the release of some neutrons and energy

MEDICAL TRACERS

Some medical diagnosis procedures use radioactive isotopes. The radiation emitted from them can be detected by a special camera which allows doctors to find out what is happening inside a patient without surgery.

ONLINE SCIENCE MAGAZINE

MEDICAL IMAGING FACES SHORTAGE OF KEY RADIOACTIVE MATERIAL

The shutdown of a Canadian nuclear reactor creates a greater than 50 percent chance of a shortage of an important medical testing marker.

Doctors use radioactive isotopes in a range of diagnostic procedures. They are chemically attached to other compounds that accumulate in a target organ. Once there, the isotopes emit radiation that specialized camera equipment picks up. The procedure is fast and, unlike biopsies and surgical procedures, non-invasive. The radiation dose is generally low and does not pose a significant health risk.

Created by the radioactive decay of molybdenum-99, technetium-99m (Tc-99m) is the most popular medical isotope, used in about 80 percent of the imaging procedures that rely on radioactive tracers. It is an unstable form of technetium-99, an isotope of the radioactive element technetium whose nucleus contains 99 protons plus neutrons.

A Complex Supply Chain

Tc-99m has a half-life – the time it takes for half of it to decay radioactively – of only six hours once it's produced. So it can't be shipped over long distances. Instead, most hospitals receive the parent isotope molybdenum-99 and let it decay into Tc-99m on site.

Complicating the situation, Mo-99 itself has a half-life of only 66 hours. That means that it can't be stockpiled. As a result, the process of transport from reactor to patient is complicated and filled with logistical, security and economic concerns.

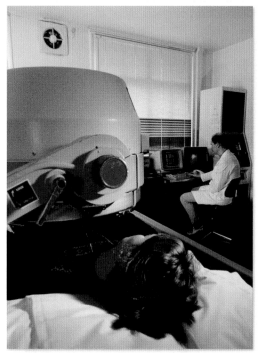

▲ **fig A:** Radioactive tracer Tc-99m is used in medical diagnosis.

From *https://www.insidescience.org/news/medical-imaging-faces-shortage-key-radioactive-material*. Dated 8th November 2016, authored by Peter Gwynne.

SCIENCE COMMUNICATION

The text is from a web-based news service connected with the American Institute of Physics.

1 (a) Discuss the tone and level of vocabulary and level of scientific detail in the story. Who is the intended audience?

(b) The full website article only had one image, and approximately four times the amount of text than in the extract shown here. How does this information alter your answer to part (a)?

2 How can you tell this article is not intended for professional physicists?

INTERPRETATION NOTE

Consider how much explanation a professional physicist would need about this topic. Compare this with the level of explanation in the article.

PHYSICS IN DETAIL

Now we will look at the physics in detail. Some of these questions will link to topics elsewhere in this book, so you may need to combine concepts from different areas of physics to work out the answers.

3 (a) What will be the difference between molybdenum-99 and technetium-99?

(b) What kind of radioactive decay must molybdenum-99 undergo to become technetium-99?

(c) The atomic number of molybdenum is 42. Write a nuclear equation for the radioactive decay that will transform molybdenum-99 into technetium-99.

4 The letter 'm' in technetium-99m means 'metastable'. This tells us that it will decay by gamma emission only.

(a) Explain why a medical tracer should be a gamma emitter for detection purposes.

(b) Explain why a medical tracer should be a gamma emitter for patient health reasons.

5 The article says that the half-life for molybdenum-99 is 66 hours, and for technetium-99m it is 6 hours.

(a) How much technetium-99 m would be produced in one week by a sample of 200 g of molybdenum-99?

(b) From a dose of 47 mg of technetium-99m, how much would be left undecayed after 24 hours?

(c) Explain why the half-life of 6 hours for Tc-99m is considered to be both usefully long, and also safely short.

PHYSICS TIP

Think about what the doctors need from the tracer, and also how to minimise the exposure for the patient.

ACTIVITY

Write a summary of five points that explain the possible impact on patients and hospitals of this shortage of technetium-99m. You should include some short-term problems, and some longer term problems.

THINKING BIGGER TIP

Think about how delays in diagnosis might affect the treatment of patients. Also think about how these delays and their effects on treatment might then affect the way the hospital works.

[Note: In questions marked with an asterisk (*), marks will be awarded for your ability to structure your answer logically, showing how the points that you make are related or follow on from each other.]

1 In an experiment to test the penetrating power of alpha particles, a Geiger–Müller counter was placed in front of the source and registered 234 counts in three minutes. A 4 cm thick block of lead was placed between the source and the GM counter and the count reduced to 26 counts in one minute. What was the background radiation level?

 A 0.433 Bq

 B 0.866 Bq

 C 1.30 Bq

 D 1.73 Bq [1]

 (Total for Question 1 = 1 mark)

2 One definition of the unified atomic mass unit has the value of 1 u as being equivalent to that of 1820 electrons. Using this definition, if an electron's mass is converted entirely into energy, how much energy would this be?

 A 7.91×10^{-6} eV

 B 0.513 MeV

 C 281 eV

 D 1 700 000 MeV [1]

 (Total for Question 2 = 1 mark)

3 The half-life of carbon-14 is 5730 years. What is the decay constant for a nucleus of carbon-14?

 A 3.84×10^{-12} s^{-1}

 B 1.21×10^{-4} s^{-1}

 C 2.52×10^{-4} s^{-1}

 D 8268 y^{-1} [1]

 (Total for Question 3 = 1 mark)

4 The diagram shows typical sources of background radiation.

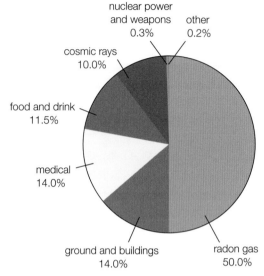

Which human-made source of radiation is less significant than 'medical' sources?

 A ground and buildings

 B cosmic rays

 C radon gas

 D nuclear power and nuclear weapons [1]

 (Total for Question 4 = 1 mark)

5 All living organisms contain ^{12}C and radioactive ^{14}C. The concentration of ^{14}C in the organism is maintained while the organism is alive, but starts to fall once death has occurred.

 (a) The count rate obtained from wood from an old Viking ship is 14.7 min^{-1} per gram of wood, after being corrected for background radiation. The corrected count rate from similar living wood is 16.5 min^{-1} per gram of wood. Calculate the age of the ship in years. ^{14}C has a half-life of 5700 years. [4]

 (b) The concentration of ^{14}C in living organisms might have been greater in the past. Explain how this would affect the age that you have calculated. [2]

 (Total for Question 5 = 6 marks)

6 Fission and fusion are both nuclear processes that release energy. About 20% of the UK's energy need is currently provided by the controlled fission of uranium. Intensive research continues to capture the energy released from the fusion of hydrogen.

(a) (i) Fission of uranium-235 takes place after the absorption of a thermal neutron.
Assume such neutrons behave as an ideal gas at a temperature of 310 K.
Show that the square root of the mean square speed of the neutrons is about $3000\ m\ s^{-1}$.
mass of neutron = 1.0087 u [3]

(ii) Complete the equation for the fission of uranium-235. [2]

$$^{235}_{92}U + ^{1}_{0}n \rightarrow ^{...}_{92}U \rightarrow ^{138}Cs + ^{96}_{37}Rb + ...^{1}_{0}n$$

(iii) Calculate the energy released in a single fission. Then determine the rate of fission necessary to maintain a power output of 2.5 GW.

	Mass/u
^{235}U	235.0439
^{138}Cs	137.9110
^{96}Rb	95.9343

[4]

(b) *(i) State the conditions for fusion and go on to explain why it has proved difficult to maintain a sustainable reaction in a practical fusion reactor. [4]

(ii) The nuclear reaction below represents the fusion of two deuterium nuclei.
Complete the equation and identify particle X. [1]

$$^{2}_{1}D + ^{2}_{1}D \rightarrow ^{3}_{1}H + ^{...}_{...}X$$

(iii) Despite the difficulties, the search for a practical fusion reactor continues.
State **two** advantages that fusion power might have compared to fission power. [2]

(Total for Question 6 = 16 marks)

7 You are planning an experiment to investigate the ability of gamma rays to penetrate lead. You are then going to analyse a set of data from such an experiment.

(a) You have a source of radiation and a detector and counter. Describe briefly a simple experiment to confirm that the source emits gamma radiation. [3]

(b) You are provided with sheets of lead and apparatus to support them safely between the source and the detector. The thickness of lead affects the count rate. Describe the measurements you would make to investigate this.
Your description should include:

- a variable you will control to make it a fair investigation
- how you will make your results as accurate as possible
- one safety precaution. [6]

(c) For gamma rays passing through lead of thickness x, the count rate A is given by:

$$A = A_0 e^{-\mu x}$$

where A_0 is the count rate when there is no lead between source and detector, and μ is a constant.
Explain why a graph of ln A against x should be a straight line. [1]

(d) The following data were obtained in such an investigation. The background count was $40\ minute^{-1}$.

x/mm	Measured count rate/minute^{-1}
0	1002
6.30	739
12.74	553
19.04	394
25.44	304
31.74	232

Process the data as appropriate and then plot a suitable graph to show that these data are consistent with $A = A_0 e^{-\mu x}$. [5]

(e) Use your graph to determine a value for the constant μ. [2]

(Total for Question 7 = 17 marks)

8 The graph shows how the binding energy per nucleon varies with nucleon number for a range of isotopes.

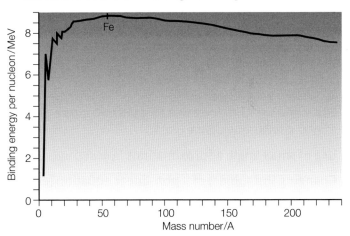

(a) Explain why an alpha particle (He^4) is more likely than any other small nucleus to be emitted from a large unstable nucleus. [2]

(b) Explain why fission reactors use isotopes such as U-235 as fuel. Your answer should include reference to the graph. [3]

(Total for Question 8 = 5 marks)

TOPIC 10 OSCILLATIONS

10A OSCILLATIONS

Oscillations are common in everyday life, and studying them can be fun! The mathematics governing these oscillations could be considered less fun, but actually is much easier than it may look when all the equations are written out together.

In this chapter we will concentrate on simple harmonic oscillations, as these are the most common and help to find useful solutions to engineering problems. We will look at how oscillations can be used for fun in everyday life, from playground toys to nightclub music.

Oscillations can also get out of control, which can be damaging or dangerous. In Angers, France, in 1850, over 220 soldiers died as the bridge they were crossing collapsed into the cold river below. It is believed that the rhythmic marching of the soldiers caused the bridge to oscillate too much and this broke the corroded cables that supported it. Following this disaster, and some other similar bridge collapses, soldiers are now trained to march out of step across bridges so as to avoid generating excessive oscillations.

We will see how such large oscillations develop and what can be done to reduce the problems they cause. These solutions range from car suspension systems to earthquake protection systems in skyscrapers.

MATHS SKILLS FOR THIS CHAPTER

- **Draw and use the slope of a tangent to a curve as a measure of rate of change** (*e.g. drawing a velocity–time graph from a displacement–time graph*)

- **Use of small angle approximations** (*e.g. in pendulum motion*)

- **Changing the subject of a non-linear equation** (*e.g. finding the gravitational field strength from a pendulum*)

- **Sketching and interpreting relationships shown graphically** (*e.g. sine and cosine functions for displacement, velocity and acceleration of an oscillator*)

- **Applying the principles of calculus** (*e.g. the connections between acceleration, velocity and displacement*)

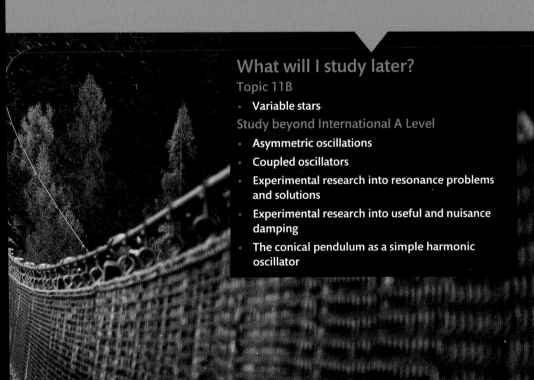

What prior knowledge do I need?

Topic 1A (Book 1: IAS)
- Speed and acceleration
- Newton's laws of motion

Topics 1B and 1C (Book 1: IAS)
- Conservation of energy and momentum

Topic 2B (Book 1: IAS)
- Elastic and plastic deformation

Topic 3A (Book 1: IAS)
- Basic wave properties

Topic 3B (Book 1: IAS)
- Wave phase

Topic 5B
- Circular motion and its connection with wave movements

What will I study in this chapter?

- The basics of oscillatory motion
- Simple harmonic motion (SHM)
- How to calculate simple harmonic motion
- How to represent SHM graphically
- Free and forced oscillations
- Resonance and damping

What will I study later?

Topic 11B
- Variable stars

Study beyond International A Level
- Asymmetric oscillations
- Coupled oscillators
- Experimental research into resonance problems and solutions
- Experimental research into useful and nuisance damping
- The conical pendulum as a simple harmonic oscillator

LEARNING OBJECTIVES

■ Describe simple harmonic motion situations, and state that the condition for simple harmonic motion is $F = -kx$.

■ Interpret a displacement–time graph for a simple harmonic oscillator.

■ Use the equations for a simple harmonic oscillator and a simple pendulum.

BOUNCING AND SWINGING

There are many things around us which **oscillate** (**figs A** and **B**). This means they have continuously repeated movements. For example, a child's swing goes backwards and forwards through the same positions over and over again. If it swings freely, it will always take the same time to complete one full swing. This is known as its time **period**, T. It follows a system of movements known as **simple harmonic motion (SHM)**. When a system is moving in SHM, a force, known as a restoring force, F, is trying to return the object to its equilibrium position, and this force is proportional to the displacement, x, from that equilibrium position.

$$F = -kx$$

where k is a constant dependent on the oscillating system in question. The negative sign in the equation shows us that the acceleration will always be towards the centre of oscillation.

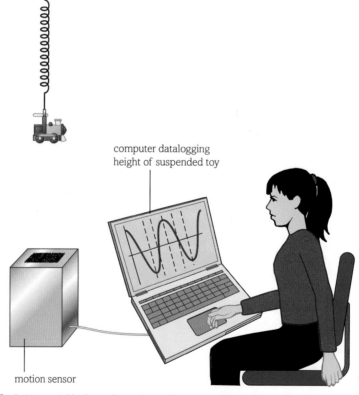

computer datalogging
height of suspended toy

motion sensor

▲ **fig A** How quickly does a bouncing spring toy move?

You will notice that the SHM definition is the same equation as in Hooke's law, and the oscillation of a mass attached to a spring is an example of SHM. We can find the equation for the period of the oscillations of a mass, m, subject to a Hooke's law restoring force:

$$T = 2\pi\sqrt{\frac{m}{k}}$$

Look at **Section 10A.2** and see if you can use the mathematics presented there to derive this equation.

WORKED EXAMPLE 1

(a) Look at **fig B**. A 500 g toy train is attached to a pole by a spring, with a spring constant of 100 N m^{-1}, and made to oscillate horizontally. What force will act on the train when it is at its amplitude position of 8 cm from equilibrium?

$$F = -kx = -(100) \times (0.08) = -8\,N$$

A force of 8 newtons will act on the train, trying to pull it back towards the equilibrium position.

(b) How fast will the train accelerate whilst at this amplitude position?

From Newton's second law:

$$a = \frac{F}{m} = \frac{-8}{0.5} = -16\,m\,s^{-2}$$

(c) What is the period of SHM oscillations for the train?

$$T = 2\pi \sqrt{\frac{m}{k}} = 2\pi \sqrt{\frac{0.5}{100}} = 0.44\,s$$

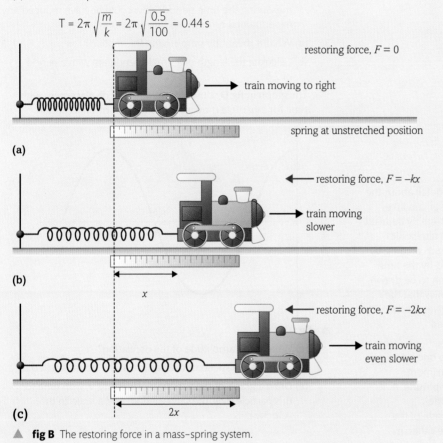

restoring force, $F = 0$

train moving to right

spring at unstretched position

(a)

restoring force, $F = -kx$

train moving slower

(b)

x

restoring force, $F = -2kx$

train moving even slower

(c)

$2x$

▲ **fig B** The restoring force in a mass–spring system.

PENDULUM DYNAMICS

Apparently, when Galileo was just 17, he was watching a hanging lantern in the cathedral in Pisa and he observed that a pendulum's time period is independent of the size of the oscillations. In fact, the period for a pendulum is given by the expression:

$$T = 2\pi \sqrt{\frac{l}{g}}$$

Therefore, the period is only dependent on the length, l, of the pendulum string, and the strength of gravity on the planet on which it has been set up.

WORKED EXAMPLE 2

The pendulum which operates a grandfather clock is 100 cm long. What is the period of oscillation?

$$T = 2\pi \sqrt{\frac{l}{g}} = 2\pi \sqrt{\frac{1.00}{9.81}} = 2.01\,s$$

PRACTICAL SKILLS

Investigating a pendulum

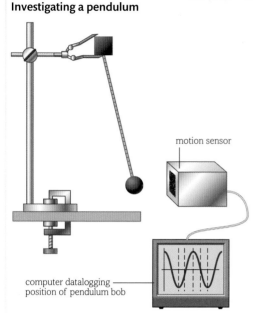

motion sensor

computer datalogging
position of pendulum bob

▲ **fig C** For small oscillations, the component of a pendulum's weight
force pushing it back to equilibrium is proportional to its horizontal
displacement from equilibrium, so it will execute SHM.

Using the set-up in **fig C**, it is easy to find out a lot of detail about the
oscillation of a pendulum. An ultrasonic position sensor can take
hundreds of readings every second to produce an extremely detailed
displacement–time graph for the pendulum bob. The ultrasonic
datalogging sensor can also measure the velocity at each reading by
making a calculation of the change in position divided by the time
between readings. This will allow a velocity–time graph to be drawn.
However, you should remember from simple mechanics that the
gradient of a displacement–time graph at any point gives the velocity
at that moment. Thus the v–t graph could be generated
mathematically from the d–t graph.

To make the pendulum oscillate, you would hold it to one side, to a
small maximum displacement, and release it. This maximum
displacement from the equilibrium position is the **amplitude**, A. It
will always be stationary, zero velocity, at the amplitude
displacement, and at the same position on the other side of the
swing. As it swings from one stationary position to the other, it
accelerates to a maximum velocity and then slows down again to a
stop. This all happens symmetrically, with the maximum velocity
occurring in the middle, at zero displacement. As the restoring force
is proportional to the displacement, this means that acceleration will
be maximum at amplitude and zero when the displacement is zero in
the middle (more details on the mathematics involved in this
experiment are given in **Section 10A.2**).

You may notice that the amplitude reduces slightly over time, but
this should not affect the time period for your pendulum. If the
period is independent of the amplitude, such oscillations are
known as isochronous.

Safety Note: Make sure the stand is clamped securely to the
bench. Use a small pendulum bob and small swings.

CHECKPOINT

SKILLS ▷ **CREATIVITY**

1. (a) Draw three free-body force diagrams showing the forces
    acting on the toy in **fig A**. The diagrams should show the
    situations when the toy is at its equilibrium position, and
    above and below the equilibrium position.

   (b) Why might this toy not follow simple harmonic motion?

2. ▷ (a) If the lantern which Galileo observed in Pisa cathedral had
    a mass of 5 kg hanging on a chain 4.4 m long, what was its
    time period?

   (b) Galileo used his own heartbeat as a stop clock when
    measuring the time period. How could he have reduced any
    experimental error in the time measurement?

3. (a) What is meant by *simple harmonic motion*?

   (b) Calculate the length of a simple pendulum with a period
    of 2.5 s.

   The graph in **fig D** shows the variation of displacement with
   time for a particle moving with simple harmonic motion.

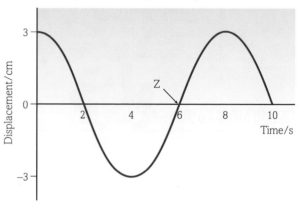

▲ **fig D**

   (c) What is the amplitude of the oscillation?

   (d) Estimate the speed of the particle at the point labelled Z.

   (e) Draw a graph of the variation of velocity, v, with time for
    this particle over the same period. Add a scale to the
    velocity axis.

SUBJECT VOCABULARY

oscillate to undertake continuously repeated movements

period the time taken for one complete oscillation

simple harmonic motion (SHM) the oscillation of a system in which
a force is continually trying to return the object to its centre position
and this force is proportional to the displacement from that centre
position

amplitude the maximum displacement from the equilibrium position

10A | 2 SHM MATHEMATICS

■ Use the equations for simple harmonic oscillators:
$a = -\omega^2 x$, $x = A \cos \omega t$, $v = -A\omega \sin \omega t$,
$a = -A\omega^2 \cos \omega t$, $\omega = 2\pi f$, and $T = \dfrac{1}{f} = \dfrac{2\pi}{\omega}$.

■ Draw and interpret displacement–time and velocity–time graphs for simple harmonic oscillators, including understanding what the gradient represents in each case.

ANGULAR VELOCITY AND SHM

For objects moving in a circle, the relationship that gave us the angular velocity is:

$$\omega = \frac{\theta}{t}$$

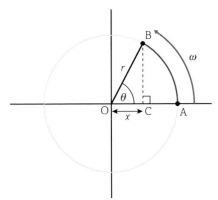

▲ **fig A** Relationships in circular motion.

In **Section 5B.1**, we also saw the relationships:

$$T = \frac{1}{f}$$

$$T = \frac{2\pi}{\omega}$$

Therefore, for an object moving in a circle:

$$\omega = 2\pi f$$

The movement of an oscillating object follows a similar pattern to that of circular motion, so all of these equations are valid in SHM. For an object performing SHM, we can determine its angular velocity, despite the fact that it may not actually be moving in a circle. Its motion is the *projection* of motion in a circle.

In **fig A**, consider the horizontally projected distance to the object from the vertical axis, and how this changes over time as the object rotates around the circle. When the object is at A, this projected distance is equal to the radius of the circle, r, but at position B this distance is shown by OC. You could calculate OC from:

$$x = r \cos \theta$$

As the object moves around the circle, its angular displacement changes according to its angular velocity, $\omega = \dfrac{\theta}{t}$. Rearranging this as $\theta = \omega t$, the displacement can be rewritten as:

$$x = r \cos(\omega t)$$

The motion of all simple harmonic oscillators can be described by an equation of this form. Indeed, all simple harmonic oscillators can be described by a sine or cosine function which gives their displacement, velocity and acceleration over time. In the case of a pendulum bob, the radius of the circle is replaced by the amplitude of the pendulum's swing. The expression for the displacement becomes:

$$x = A \cos \omega t \qquad \text{because at } t = 0 \;\; x = A$$

The pendulum that operates a grandfather clock is released from its amplitude position which is 10 cm from the middle. It swings completely through one cycle every two seconds. What is its angular velocity, and where will it be after 174.5 seconds?

$$A = 10\,\text{cm} = 0.1\,\text{m}; \; T = 2\,\text{s}$$

$$\omega = \frac{2\pi}{T} = \frac{2\pi}{2}$$

$$\therefore \quad \omega = 3.14\,\text{rad s}^{-1}$$

Releasing from maximum displacement means

$$x = A \cos \omega t$$

$$x = 0.1 \cos(3.14 \times 174.5)$$

$$= 0.1 \cos(547.93)$$

$$= 0.1 \times 0.274$$

$$\therefore \quad x = 0.027\,\text{m}$$

SHM GRAPHS

When an object is moving with SHM the restoring force is proportional to the displacement, $F = -kx$. As we can describe the position using $x = A \cos \omega t$, this gives the equation for the force over time as:

$$F = -kA \cos \omega t$$

Not all SHM oscillations involve a spring, so here k refers to some constant relevant to the oscillator set-up. From Newton's second law, we can also show that:

$$ma = -kx$$

so: $\quad a = -\dfrac{kx}{m} = -\dfrac{k}{m} A \cos \omega t$

From this we can see that the acceleration and displacement in SHM have the same form, but the acceleration acts in the *opposite* direction to the displacement. When the displacement is zero, so is the acceleration. And when x is at its maximum value, the acceleration is also at its maximum value (**fig B**).

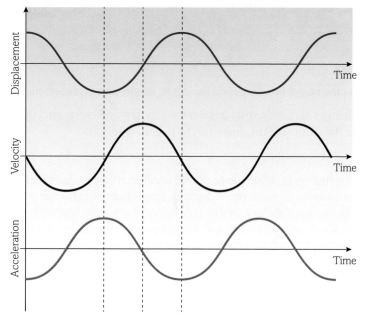

▲ **fig B** The changes in position, velocity and acceleration over time for a simple harmonic oscillator. The displacement can be described using either sine or cosine (depending upon whether the oscillations are taken to start from the centre or at amplitude). Velocity and acceleration are then also sine or cosine as appropriate.

We can calculate the velocity of an oscillator at any moment from the gradient of the displacement–time graph. For a function, $x = \sin\theta$, the derivative function, $\dfrac{\mathrm{d}x}{\mathrm{d}t}$, which tells us the gradient at each point, is $\dfrac{\mathrm{d}x}{\mathrm{d}t} = \cos\theta$. For $x = \cos\theta$, the derivative function is $\dfrac{\mathrm{d}x}{\mathrm{d}t} = -\sin\theta$.   (Remember, θ is a function of time.)

If $x = A\cos\omega t$, then the change in displacement with time is given by:

$$\frac{\mathrm{d}x}{\mathrm{d}t} = v = -A\omega\sin\omega t$$

The change in velocity with time, the acceleration is:

$$\frac{\mathrm{d}^2x}{\mathrm{d}t^2} = \frac{\mathrm{d}v}{\mathrm{d}t} = a = -A\omega^2\cos\omega t$$

or: $a = -\omega^2 x$

If we know the mass of the oscillating object, and the angular velocity of the oscillation, we can work out the restoring force constant, k, in any SHM.

$$a = -\frac{kx}{m} = -\omega^2 x$$

$\therefore$ $k = \omega^2 m$

LEARNING TIP

The maximum values of sin and cos are both 1. So, the maximum values of velocity and acceleration are:

$$v_{max} = -A\omega_{amax}$$

$$a_{max} = -A\omega^2$$

WORKED EXAMPLE 2

A science museum has a giant demonstration tuning fork. The end of each prong vibrates in simple harmonic motion with a time period of 1.20 s and starts vibrating from an amplitude of 80 cm. Calculate the displacement, velocity and acceleration of a prong after 5.0 seconds.

$A = 0.8\,\mathrm{m}$; $T = 1.20\,\mathrm{s}$

$\omega = \dfrac{2\pi}{T} = \dfrac{2\pi}{1.20} = 5.24\ \mathrm{rad\ s^{-1}}$

$x = A\cos\omega t = 0.80 \times \cos(5.24 \times 5.0) = 0.386\,\mathrm{m}$

$v = -A\omega\sin\omega t = -0.8 \times 5.24 \times \sin(5.24 \times 5.0) = -3.67\ \mathrm{m\ s^{-1}}$

$a = -A\omega^2\cos\omega t = -\omega^2 x = -(5.24)^2 \times (0.386) = -10.6\ \mathrm{m\ s^{-2}}$

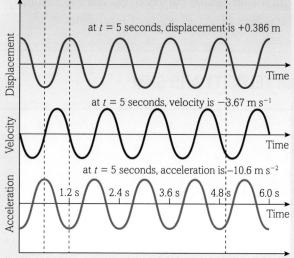

▲ **fig C** SHM graphs for a giant tuning fork.

CHECKPOINT

1. What is the angular velocity of a child's swing which completes 42 swings every minute?

SKILLS   CREATIVITY

2. ▶ Comment on the rounding errors in the worked example about the giant tuning fork above. Estimate the percentage errors generated.

SKILLS   ANALYSIS

3. ▶ The toy train in **fig B** of **Section 10A.1** is released from an amplitude position of 5 cm. Calculate its position, velocity and acceleration after 2.8 seconds.

4. A grandfather clock has a 2-metre-long pendulum to keep time. The owner set it at noon by setting the hands and starting the pendulum from an amplitude of 10 cm. Calculate the position, velocity and acceleration of the pendulum bob at 6 seconds after noon.

10A 3 SHM ENERGY

CONSERVATION OF ENERGY

An oscillator is a closed system which means that it cannot gain or lose energy, unless there is an external influence. For example, a pendulum under water will not continue to swing for very long. This is because the viscous drag on the pendulum from the water provides an external force, which can do work to remove energy from the pendulum system. A child on a swing can continue to oscillate, despite air resistance removing energy, if there is also a parent pushing repeatedly (doing work) to put energy back into the system. Such damped or forced oscillations will be dealt with in the next sections. Here we will only consider a closed system in which no energy moves in or out of the system during its oscillations.

ENERGY TRANSFERS DURING OSCILLATION

The kinetic energy of a pendulum varies during its swinging motion. At each end of the swing, the bob is stationary for an instant, meaning that it has zero kinetic energy at these points. The kinetic energy then increases steadily until it reaches a maximum, when the bob passes through the central position and is moving with its maximum speed. The kinetic energy then decreases to reach zero at the other amplitude position. This cycle repeats continuously. However, once swinging, the pendulum system cannot gain or lose energy. The varying kinetic energy must be transferring back and forth into another store. In this example, the other store is gravitational potential energy as the pendulum rises to a maximum height, where it has maximum GPE at each end of the swing and drops to minimum GPE as it passes through the lowest point (the central position). These energy changes are shown in **fig A**.

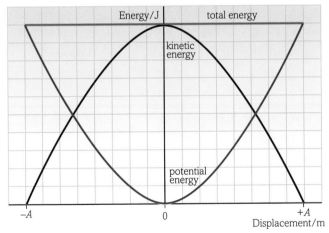

▲ **fig A** The energy in a simple harmonic oscillator is constantly being transferred from kinetic to potential and back again, whilst the total remains constant as the sum of the kinetic and potential energies at each moment.

Fig A could also apply to a mass bouncing on a spring. When stretched a little before release, the spring has stored elastic potential energy. On release, the spring accelerates the mass back towards the equilibrium position, where it has maximum kinetic energy and zero potential. This kinetic energy is then transferred into elastic potential as the spring squashes in the other displacement direction. The transfers of potential and kinetic energies continues, and if no energy can be lost from the system it would oscillate with the same amplitude forever.

The above discussions are exclusively for a system that cannot lose any energy. The closest we can get to this system would be a pendulum, or mass on a spring, in a vacuum. These would be almost free from energy losses, but air resistance will act in most real situations.

▲ **fig B** Converting potential energy into kinetic energy can be fun!

EXAM HINT

When working through a long multi-step calculation, always keep using the exact figures and only round off the final answer.

WORKED EXAMPLE 1

A forest playground has a tyre hanging from a tree branch. The tyre behaves like a pendulum, with a rope of 4.0 metres length, and its mass is 15 kg. A child of mass 45 kg swings on the tyre by pulling it 3.0 metres to one side (amplitude) and leaping on. What is the maximum height that the tyre reaches above its equilibrium position?

$$T = 2\pi\sqrt{\frac{l}{g}} = 2\pi\sqrt{\frac{4.0}{9.81}} = 4.0\,s$$

$$\omega = \frac{2\pi}{T} = 1.6\,rad\,s^{-1}$$

$$v = -A\omega\sin\omega t$$

At maximum velocity, $\sin\omega t$ will equal 1, so:

$$v_{max} = A\omega = 3.0 \times 1.6 = 4.8\,m\,s^{-1}$$

Maximum kinetic energy:

$$E_{kmax} = \tfrac{1}{2}mv_{max}^2 = \tfrac{1}{2} \times (45 + 15) \times (4.8)^2 = 691\,J$$

Maximum potential energy:

$$E_{pmax} = mgh_{max} = E_{kmax} = 691\,J$$

$$\therefore \quad h_{max} = \frac{691}{mg}$$

$$h_{max} = 1.2\,m$$

WORKED EXAMPLE 2

The toy train in **fig B** of **Section 10A.1** is released from an amplitude position of 5 cm. Calculate its maximum velocity. What assumption have you made?

$$A = 0.05\,m \quad k = 100\,N\,m^{-1} \quad m = 0.500\,kg$$

Maximum potential energy:

$$E_{pmax} = \tfrac{1}{2}F\Delta x = \tfrac{1}{2}k(\Delta x)^2$$

$$= \tfrac{1}{2} \times 100 \times (0.05)^2$$

$$\therefore E_{pmax} = 0.125\,J$$

Maximum kinetic energy:

$$E_{kmax} = E_{pmax} = \tfrac{1}{2}mv_{max}^2 = 0.125\,J$$

$$v_{max} = \sqrt{\frac{2 \times 0.125}{0.5}}$$

$$\therefore v_{max} = 0.71\,m\,s^{-1}$$

We have assumed that there are no energy losses in the system, so all the elastic potential energy is transferred into kinetic energy.

▲ **fig C** A bungee jumper

CHECKPOINT

SKILLS   CRITICAL THINKING, CONTINUOUS LEARNING

1. ▶ A pendulum bob has a mass of 0.6 kg and a time period of 4 seconds. It is released from an amplitude position of 5 cm. What is its kinetic energy after 8 seconds? What is its maximum kinetic energy?

2. ▶ If no energy were lost, a bungee jumper (as in **fig C**) would continue to oscillate up and down forever. Describe the energy changes the bungee jumper would undergo. Discuss kinetic energy, different types of potential energy, and the total energy. Conclude by explaining why in real life this does not happen.

3. ▶ Suggest how energy might be lost from a pendulum swinging in a vacuum.

4. Pendulum dynamics assumes that the oscillations are of a small amplitude compared with the length of the pendulum. The worked example of the forest tyre pendulum at the top of this page does not meet this assumption. Use a geometrical method to calculate the vertical height gained when the child holds the tyre at the 3 m horizontal amplitude quoted. Compare your answer with the 1.2 m calculated in the worked example, and calculate the percentage difference between the two answers.

10A 4 RESONANCE AND DAMPING

SPECIFICATION
REFERENCE

5.5.148	5.5.149	5.5.151
5.5.152	5.5.153	CP16

CP16 LAB BOOK PAGE 67

LEARNING OBJECTIVES

- Explain the distinction between free and forced oscillations.
- Define the phenomenon of resonance.
- Describe how damping affects oscillations, how damping can be caused, and its uses and implications.

FREE AND FORCED OSCILLATIONS

Releasing a pendulum from its maximum amplitude and letting it swing freely (preferably in a vacuum) is an example of **free oscillation**. The situation is set up for a continuous exchange of potential and kinetic energy, caused by a restoring force which is proportional to the displacement. Any oscillating system has a **natural frequency** – the frequency at which it naturally chooses to oscillate when left alone.

However, oscillators can be forced to behave in a different way to their natural motion. If, as a pendulum swings one way, you push in the opposite direction, it turns back. By repeated applications of forces from your hand in different directions, you could force the pendulum to oscillate at some other frequency. This would then be **forced oscillation** – not SHM – and the frequency at which you were causing it to swing at would be your **driving frequency**.

Forcing oscillations involves adding energy to a system whilst it oscillates. Unless this is done at the natural frequency, the system is unlikely to undergo SHM and will dissipate the energy quite quickly. This is what happens if you push a child on a swing at the wrong moment in the oscillation.

RESONANCE

If a system or object is forced to vibrate at its natural frequency, it will absorb more and more energy, resulting in very large amplitude oscillations. This is **resonance**. If your washing machine is very noisy during one part of the wash cycle, the motor will be spinning at a certain frequency. This drives vibrations in the machine at a frequency that is the natural frequency of one of the panels, which then vibrates with very large amplitude, making a loud noise. During the rest of the wash, the motor's rotation generates vibrations at other frequencies that do not match with the natural frequency of any part of the machine, and it is much quieter.

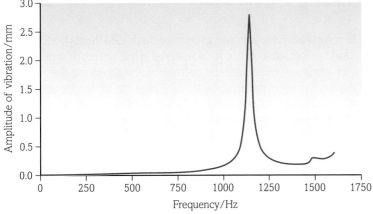

▲ **fig A** Graphical illustration of how the driving frequency in a washing machine affects the oscillation amplitude of its side panel. At all frequencies here, the amplitude of forced vibration is the same, and relatively small. When the motor's vibration reaches about 1100 Hz, the graph peaks, showing that the side panel vibrates significantly, which will be very noisy.

Investigating resonance

The set-up in **fig B** allows us to monitor the amplitude of vibrations of a mass on a spring, as we force it to vibrate at different frequencies. By monitoring for a very large amplitude of vibration, we can determine the resonant (or natural) frequency for the set-up.

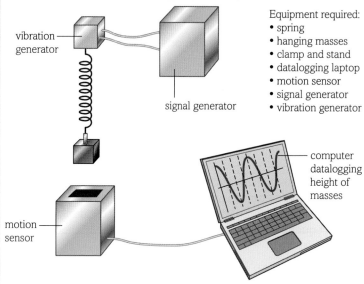

Equipment required:
- spring
- hanging masses
- clamp and stand
- datalogging laptop
- motion sensor
- signal generator
- vibration generator

▲ **fig B** Forced oscillations can be set up at a desired driving frequency.

From previously, we know that a mass oscillating on a spring follows the equation for its time period:

$$T = 2\pi \sqrt{\frac{m}{k}}$$

The time period will be the reciprocal of the natural frequency, f, and this is also the resonant frequency:

$$f = \frac{1}{2\pi} \sqrt{\frac{k}{m}}$$

Thus, by finding the resonant frequency, from monitoring the amplitude of forced oscillations, we could determine an unknown mass that is hanging on the spring:

$$m = \frac{k}{4\pi^2 f^2}$$

The spring constant, k, could be found by measuring the resonant frequency for a known mass and rearranging the equation to find k:

$$k = 4\pi^2 f^2 m$$

> ⚠ Safety Note: Use a stand clamped to the bench to hold the vibration generator. Use a small mass and keep the vibrations small.

DAMPED OSCILLATIONS

Damped oscillations suffer a loss in energy in each oscillation and this reduces the amplitude over time (**fig C**).

If a system is performing SHM at its natural frequency, its energy may be dissipated through a friction force acting on the system, or the plastic deformation of a ductile material in the system. If a pendulum is left to swing without interference, its amplitude will constantly decrease with each swing due to air resistance and internal stresses within the flexing material of the string. These effects can be amplified if we attach a small sail to catch the air. This artificial increasing of the air resistance is an example of **damping** (or in fact an increase in the damping, as there would already be a tiny air resistance force on the pendulum). Note that although the amplitude decreases, the period remains constant throughout.

The amount of damping may vary, which will change how quickly the amplitude is reduced. If the oscillator completes several oscillations, the amplitude will decrease exponentially. This is known as underdamping (sometimes called light damping). Swinging the pendulum in a bowl of water will make its amplitude of oscillation drop very rapidly, and it might not even complete one cycle. This is known as overdamping.

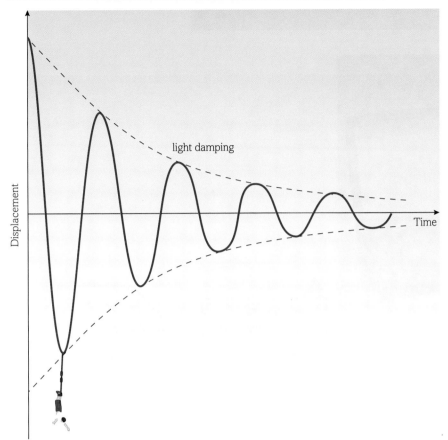

fig C Damping reduces oscillation amplitude. A bungee jumper's amplitude decreases because elastic stresses in the rubber rope dissipate energy as heat, and air resistance removes kinetic energy. Note that the dissipation of sound energy by screaming will not reduce the amplitude as this does not come from the oscillation energy!

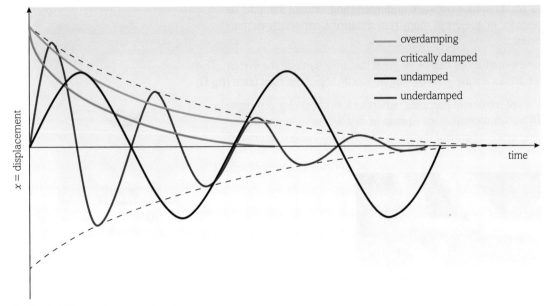

fig D Different degrees of damping.

If the damping is such that the oscillator returns to its equilibrium position in the quickest possible time, without going past that position (known as overshooting), then it is **critically damped**. Overdamping is useful in many situations, such as car suspension systems, but too much overdamping could cause the oscillator to take a very long time to return to its equilibrium position. You can appreciate this if you imagine a pendulum swung through treacle – an extreme example of overdamping.

Investigating damping

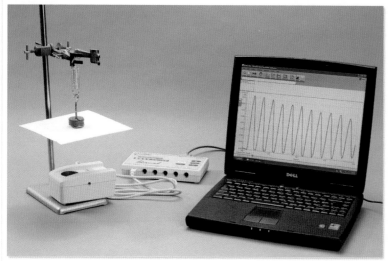

▲ **fig E** How does damping affect oscillations?

In **fig E**, you can see how to investigate the effects of more or less damping. The card sail will act as an air resistance, and altering its size will change its damping effect, making the amplitude decrease more. Can you make a card sail that provides critical damping?

! Safety Note: Make sure the spring is held securely in the clamp and only hang small masses on the spring.

RESONANCE PROBLEMS AND DAMPING SOLUTIONS

Damping solutions to real problems, such as the noisy washing machine, can be designed into new products and structures (**fig F**). Objects usually oscillate with large amplitudes at their natural frequency, so this is the frequency at which these engineering solutions are most often aimed. Creating damping systems that are a good balance between underdamping, critical damping and severe overdamping is a tough problem for engineers in many real situations, especially as many objects have several natural frequencies.

In some cases, the amplitude of resonant oscillations can be so great that they damage the system. This is what happens when an opera singer breaks a wine glass by singing loudly at just the right pitch (**fig G**).

For some oscillators (e.g. clocks) we want minimum damping; whereas for others (e.g. a mountain bike shock-absorbing system (**fig H**)) we want oscillations to stop as fast as possible. To make this happen requires an exact damping system ('critical damping') so that the system returns to equilibrium without overshooting.

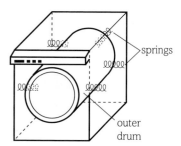

▲ **fig F** Damping in a washing machine lessens the effects of resonant vibrations.

▲ **fig G** The resonance of a glass can damage it.

DID YOU KNOW?
Shattering a wine glass by singing is a classic example of resonance (as explored in the 2005 Discovery Channel television show *MythBusters*). But it is actually only possible if the glass already contains microscopic defects where cracks can start.

▲ **fig H** Mountain bike suspension has springs to accept some movement, and shock absorbers to critically dampen any oscillations.

PRACTICAL SKILLS

Investigating damping solutions to design problems

A signal generator can be connected to a vibration generator to produce driving forces with a changeable frequency. This can then be used to investigate the effects of damping on oscillations. You could set up a model of an earthquake shaking a building and see how dampers could be attached to the model, either internally or externally, to reduce the damage caused by resonance.

▲ **fig I** Vibration testing of a model building to simulate earthquakes.

Professional research institutions will use large scale models on shaking platforms to test new designs for skyscraper protection. In a school laboratory, a structure made of straws or struts of wood could be tested with foam padding dampers and a vibration generator.

As we saw in **Book 1, Section 2B**, climbing ropes need to be made of a material which will stretch slightly, without breaking, to stop a climber from falling. This is arranged through the construction of climbing ropes. There are many strands woven together, and if you follow one individual strand, it forms a spiral. The climber hanging on the rope can act like a mass hanging on a spring. However, the material chosen must be stiff enough. If a climber fell whilst using a bungee cord as a safety rope, they would bounce many times before stopping. It would not be uncomfortable, but could well cause the climber to hit the rock face (several times!). So, the stiffness and construction of the rope must be a perfect combination of stretchy which avoids a sudden, painful stop to a fall, but still stiff enough to bring the climber to a halt. After a fall, a climbing rope has been permanently damaged and must be replaced. The rope material is ductile and suffers a plastic deformation which acts to reduce the amplitude of oscillations. A less stiff material could avoid replacement but would not damp the oscillations safely. In effect, the plastic deformation of the ductile rope material provides critical damping. Potential oscillations are stopped before they start. During an earthquake, steel framed buildings can absorb some of the vibration energy because steel is ductile, so its plastic deformation damps the oscillations.

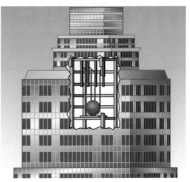

▲ **fig J** In the Taipei 101 skyscraper, a giant steel pendulum, with a mass of 660 tonnes, is tethered by giant shock absorbers. It is the world's largest wind vibration damper.

CHECKPOINT

1. Give a real life example of:
   (a) forced oscillations   (b) free oscillations
   (c) damping   (d) resonance.

   SKILLS REASONING/ARGUMENTATION

2. ▶ Explain why a car which is accelerating may have an annoying rattle of the dashboard for part of the time.

   SKILLS PRODUCTIVITY

3. ▶ Describe, using scientific terminology, how the problem in question 2 could be overcome.

4. Explain the differences between overdamping, underdamping and critical damping.

5. Explain how a damper between girders in a skyscraper could reduce damage caused by an earthquake.

   SKILLS ADAPTIVE LEARNING

6. ▶ Explain how some dance music that has an inaudible bass frequency of between 5 and 10 Hz can get people dancing.

SUBJECT VOCABULARY

free oscillation the oscillation of a system, free from the influence of any forces from outside the system

natural frequency the frequency of oscillation that a system will take if it undergoes free oscillations

forced oscillation the oscillation of a system under the influence of an external (usually repeatedly applied) force

driving frequency the frequency of an external force applied to a system undergoing forced oscillations

resonance very large amplitude oscillations that occur when a driving frequency matches the natural frequency of a system

damped oscillations work is done on the damping system and energy is dissipated in the damping system with each oscillation, so the amplitude of oscillations decreases

damping the material, or system, causing energy loss during each damped oscillation

critical damping when damping is such that the oscillator returns to its equilibrium position in the quickest possible time, without going past that position

TAIPEI 101

SKILLS CRITICAL THINKING, PROBLEM SOLVING, INTERPRETATION, CONTINUOUS LEARNING, ADAPTIVE LEARNING, PRODUCTIVITY, COMMUNICATION

Taipei 101 is a 100-storey skyscraper. It was designed to be safe in a number of extreme conditions, including very strong winds and earthquakes. Near the top of the building, there is a system to reduce the effects of resonance from these conditions.

ONLINE INFORMATION

TUNED MASS DAMPER (TMD)

The complete name of the TAIPEI 101 wind damper is the tuned mass damper (TMD). The TMD has been specifically designed as a passive damper system and is positioned at the center of the tower between the 87th and 92nd floors. Its main purpose is to reduce the swaying of the tower during strong winds and eliminate any resulting discomfort experienced by anyone within the building. Unlike conventional damper systems that are usually hidden from public view, special functional and aesthetic considerations have been made for the TAIPEI 101 wind damper so that visitors can take a look at the entire wind damper system and see how it operates at the Observatory.

▲ **fig A** The viewing gallery of the Taipei 101 Tuned Mass Damper.

From *https://www.taipei-101.com.tw/en/observatory-damper.aspx*

THE 728-TON TMD

To achieve stability and lessen the impact of violent motion, a gigantic tuned mass damper was designed. The damper consists of a steel sphere 18 feet across and weighing 728 ton, suspended from the 92nd to the 87th floor. Acting like a giant pendulum, the massive steel ball sways to counteract the building's movement caused by strong gusts of wind. Eight steel cables form a sling to support the ball, while eight viscous dampers act like shock absorbers when the sphere shifts. The ball can move 5 ft. in any direction and reduce sways by 40 percent. Two additional tuned mass dampers, each weighing 7 tons, installed at the tip of the spire provide additional protection against strong wind loads.

The engineers were so proud of their creation that they made the damper publicly visible from an indoor observatory located inside the tower, where recorded voice tours and informative displays explain to visitors how the thing works. During particularly windy days one can see the damper in action.

From *http://www.amusingplanet.com/2014/08/the-728-ton-tuned-mass-damper-of-taipei.html*

SCIENCE COMMUNICATION

The two extracts are from the website of the Taipei 101 building itself, and then a website called AmusingPlanet.com which tells stories about amazing things around the world.

1 (a) Describe the level of scientific language used in the extracts. What is the intended audience for each?

 (b) Explain what shows us that the second quote from the AmusingPlanet website is aimed at readers in the USA.

INTERPRETATION NOTE

Consider how measurements are presented to the reader.

PHYSICS IN DETAIL

Now we will look at the physics in detail. Some of these questions will link to topics elsewhere in this book, so you may need to combine concepts from different areas of physics to work out the answers.

2 (a) What physics phenomenon could allow the wind to cause surprisingly large movements of the skyscraper?

 (b) Why would the skyscraper suffer different variations of this problem with different wind speeds?

 (c) Why are large movements of the building a problem?

 (d) Explain how an earthquake could also cause a similar problem, including similarities and differences with swaying caused by the wind.

3 (a) How does this mass damper stop the building from swaying too much? Explain the physics principles behind it.

 (b) Why is the damper positioned so high up the building, around the 90th floor?

4 (a) Calculate the density of the steel used in the giant sphere. The unit 1 foot is approximately 30.8 cm, and 1 ton is approximately 1.1 metric tonnes.

 (b) The sphere hangs from eight cables that are each 42 m in length. Assuming it acts as a simple pendulum, calculate its natural frequency.

5 The giant steel sphere has its movements restricted by large shock absorbers. These are hydraulic pistons which provide large forces opposing any movements. Why are these shock absorbers necessary?

PHYSICS TIP

Think about how the sphere might move if it were free to swing.

ACTIVITY

The owners of the Taipei 101 building would like to add an information board specially for young visitors in the Observatory viewing area. Design a storyboard of diagrams that could be drawn out like a comic to show young children how the tuned mass damper reduces the swaying of the building. You may wish to look up the character 'Damper Baby' and use them in your storyboard.

THINKING BIGGER TIP

You do not need to draw out the comic in detail, just plan the picture sequence and what each picture would show, and note how this explains the physics of each step.

1 What is the length of a simple pendulum that has a time period of 1.24 s on Earth?

  A 0.00397 m

  B 0.308 m

  C 0.382 m

  D 2.4 m [1]

(Total for Question 1 = 1 mark)

2 If an oscillator is performing simple harmonic motion:

  A Its acceleration is proportional to its velocity and in the same direction.

  B Its acceleration is proportional to its displacement and in the opposite direction.

  C Its velocity is proportional to its displacement and in the same direction.

  D Its acceleration is inversely proportional to its displacement and in the opposite direction. [1]

(Total for Question 2 = 1 mark)

3 A mass suspended on a spring can undergo simple harmonic motion because:

  A The spring obeys Hooke's law that $F = -kx$.

  B The weight of the mass will be $W = mag$.

  C The spring's elastic force will form a Newton's third law pair with the weight of the mass.

  D The spring will act as a damper. [1]

(Total for Question 3 = 1 mark)

4 Which of these is a correct equation for the acceleration in simple harmonic motion?

  A $a = -\omega^2 x$

  B $a = -\omega x^2$

  C $a = -A\omega \cos\omega t$

  D $a = \omega^2 x$ [1]

(Total for Question 4 = 1 mark)

5 (a) Define a simple harmonic motion. [2]

  (b) The graph shows the variation in water level displacement with time for the water in a harbour. The water level displacement varies with simple harmonic motion.

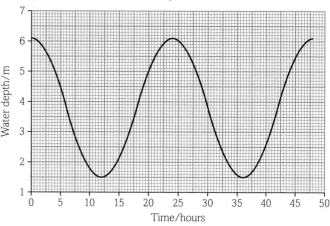

  (i) Use the graph to calculate the amplitude and the time period of the variation in the water level displacement. [2]

  (ii) Show that the maximum rate of change of water level displacement is about 0.6 m hour⁻¹. [3]

  (iii) On a copy of the graph shown in part (b), sketch how the rate of change of water level displacement varies with time for the interval 0–30 hours. The variation in water level displacement with time has been drawn for you. You need not add any numerical values to the y-axis. [2]

(Total for Question 5 = 9 marks)

6 A student makes the 'ruler piano' shown.

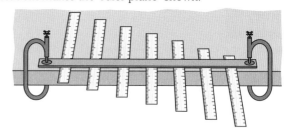

One end of each ruler is held flat on the desk whilst the other end is set into oscillation.
Each ruler oscillates at a different frequency. Some of the rulers produce an audible sound.

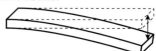

(a) State the condition for an oscillation to be simple harmonic. [2]

(b) The end of one ruler moves through 5.0 cm from one extreme position to the other, and makes 10 complete oscillations in 4.5 s.
Calculate the maximum velocity of this end. [3]

(c) A standing wave is set up on each oscillating ruler. Explain why each length of ruler oscillates at a different frequency. [3]

(Total for Question 6 = 8 marks)

7 A bar magnet is suspended by a thread attached to a wooden support. The bar magnet hangs horizontally as shown and lines up with the Earth's magnetic field.

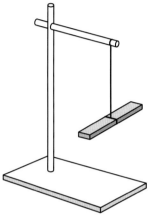

(a) State why the support should be made of wood and not steel. [1]

(b) The magnet is rotated horizontally about its centre through approximately 20° from its equilibrium position. When it is released it oscillates in a horizontal plane about the string.

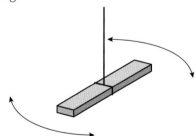

Describe how you would measure the period of these oscillations as accurately as possible. [3]

(c) You are told that these oscillations are lightly damped. State what you would observe if the oscillations were heavily damped. [1]

(d) A large coil of wire is now placed vertically around the centre of the magnet.

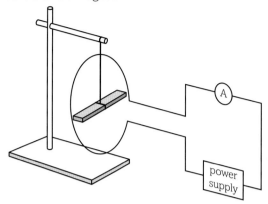

When current is passed through the coil it produces a magnetic field **in the same direction** as the Earth's magnetic field. When the magnet is again rotated horizontally, it oscillates at an increased frequency.

(i) A student thinks that the period T of the oscillations is related to the current in the coil I by

$$\frac{1}{T^2} = kI$$

where k is a constant.
Explain why this suggests that a graph of $\frac{1}{T^2}$ against I will produce a straight line through the origin. [2]

(ii) The student carries out an experiment to measure T as I is varied. He obtains the following data.

I/A	Mean T/s
0	1.230
1.00	0.827
2.00	0.673
3.00	0.581
4.02	0.520
5.01	0.475

Plot a graph that would test the relationship:

$$\frac{1}{T^2} = kI$$

Draw your own table to show any processed data. [5]

(iii) The student's teacher suggests that the equation

$$\frac{1}{T^2} = kI + b$$

is a better mathematical model for the data.
Explain why his teacher is right. [2]

(Total for Question 7 = 14 marks)

8 Explain how damping and the plastic deformation of ductile materials affects oscillations. Include explanations of overdamping, underdamping and critical damping. [6]

(Total for Question 8 = 6 marks)

TOPIC 11 ASTROPHYSICS AND COSMOLOGY

11A GRAVITATIONAL FIELDS

Gravity is one of the most obvious scientific phenomena and we have studied it for a very long time. When you drop a pen you can see gravity and its effects clearly and we know how to utilise these in very precise ways.

In November 2014, the European Space Agency landed a probe on the comet 67P/Churyumov–Gerasimenko. This may not sound remarkable. We have landed probes on the Moon, Mars, other planets and moons of other planets. However, the size and mass of this comet means that the gravitational field is so weak that it would be difficult to create a centripetal force of attraction on the probe which would allow orbiting. The ESA had to ensure that the lander was travelling at less than one metre per second, and even then the lander bounced significantly.

However, gravity remains one of the least well-understood areas of physics. There are a number of experiments currently ongoing to try and observe some of the theoretical explanations of how gravity works on a fundamental level. As yet, none have been conclusive, but as their sensitivity improves, scientists hope that we will be able to collect enough experimental evidence to confirm how gravity works. Gravity waves, quantum gravity and so-called strong gravity are current areas of interest to cutting-edge physicists.

MATHS SKILLS FOR THIS CHAPTER

- **Solving algebraic equations** (*e.g. finding a satellite's orbital height above the surface of the Earth*)

- **Changing the subject of a non-linear equation** (*e.g. finding the separation of objects using Newton's universal law of gravitation*)

- **Sketching and interpreting relationships shown graphically** (*e.g. comparing sketch graphs of gravitational field strength and potential*)

- **Distinguishing between average and instantaneous rates of change** (*e.g. gravitational field strength as the rate of change of potential with distance*)

What prior knowledge do I need?

Topic 1A (Book 1: IAS)
- Weight and gravity
- Newton's laws of motion

Topic 1B (Book 1: IAS)
- Gravitational potential energy

Topic 5B
- Circular motion

Topic 6A
- Coulomb's law
- Electric fields and electric potential

What will I study in this chapter?
- How gravitational forces follow an inverse square relationship
- What is meant by a gravitational field
- How to calculate the strength of gravitational fields
- Gravitational potential
- Comparisons of electric and gravitational fields

What will I study later?

Topic 11B
- How gravity leads to the development of stars
- The production of elements in stellar nuclear fusion
- The gravitational field strength at the surface of a black hole
- The influence of gravity on the development of galaxies and the Universe as a whole
- Dark matter and its gravitational effects
- Gravitational lensing as evidence for black holes

LEARNING OBJECTIVES

■ Use the equation $F = \dfrac{Gm_1m_2}{r^2}$.

■ Define a gravitational field, and gravitational field strength.

■ Be able to apply Newton's laws of motion and universal gravitation to orbital motion.

THE ATTRACTION OF GRAVITY

The current theories about the formation of the galaxies, stars and planets are all driven by one fundamental underlying force: gravity. Every particle with mass attracts every other particle with mass. So, if the Big Bang spread tiny particles across space, they would all attract each other. The acceleration generated by these tiny forces might be exceedingly small, but the Universe has plenty of time. Slowly but surely, the particles move towards each other and clump together. This effect is greater for particles nearer to each other, as they attract more strongly than those separated by larger distances. These clumps of matter will continue to attract other nearby particles or lumps that have formed, and continue to **accrete** (come together under the influence of gravity) into larger and larger bodies of material. This collection of matter might become a planet, or if enough material gets packed together densely enough, nuclear fusion may start and it becomes a star.

▲ **fig A** Gravity drives the formation of stars – here from clouds of gas in the Carina nebula.

Like electric fields created by charged particles, a massive particle will also generate a radial gravitational field around itself. A particle that has mass will experience a force when it is in a **gravitational field**. Unlike electric fields, gravity is always attractive.

The force that a body will experience is the strength of the gravitational field (g) multiplied by the amount of mass (m), as given by the equation:

$$F = mg$$

From this force equation, we can also see how quickly a massive particle would accelerate. From Newton's second law we know that $F = ma$, so we can equate the two equations:

$$F = mg = ma$$

So: $a = \dfrac{mg}{m} = g$

Therefore, what we have previously referred to as the *acceleration due to gravity* is the same as the *gravitational field strength*, g. On Earth, these are $9.81\ \text{m s}^{-2}$ or $9.81\ \text{N kg}^{-1}$.

WORKED EXAMPLE 1

What force will a two-tonne elephant feel when it is in the Moon's surface gravitational field, which has a strength of $1.62\ \text{N kg}^{-1}$?

$$F = mg$$
$$= 2000 \times 1.62$$
$$F = 3240\ \text{N}$$

▲ **fig B** The Moon's weaker gravitational field makes Earth-developed muscles seem extra strong.

GRAVITATIONAL FORCES

▲ **fig C** As with much of modern science, the law of the force of gravity between two objects was first published by Sir Isaac Newton.

Newton was the first scientist to publish the equation that gives us the gravitational force between two masses, m_1 and m_2, which are separated by a distance, r, between their centres of gravity:

$$F = \frac{Gm_1m_2}{r^2}$$

where G is the gravitational constant, $G = 6.67 \times 10^{-11}\,\text{N}\,\text{m}^2\,\text{kg}^{-2}$.

WORKED EXAMPLE 2

(a) What force will exist between two neutrons in the Crab Nebula which are two metres apart?

$$F = \frac{Gm_1m_2}{r^2}$$
$$= \frac{6.67 \times 10^{-11} \times 1.67 \times 10^{-27} \times 1.67 \times 10^{-27}}{2^2}$$
$$F = 4.65 \times 10^{-65}\,\text{N}$$

(b) How quickly will each accelerate towards the other (ignoring the effects of all other particles)?

$$a = \frac{F}{m}$$
$$= \frac{4.65 \times 10^{-65}}{1.67 \times 10^{-27}}$$
$$a = 2.78 \times 10^{-38}\,\text{m}\,\text{s}^{-2}$$

(c) Why is it difficult to calculate how long they would take to reach each other?

The acceleration is inversely proportional to the square of the distance between them. Therefore, the acceleration is constantly increasing as they get closer together. Our usual equations of motion require **uniform acceleration**.

WEIGHING THE EARTH

Newton thought about how the Moon knows that the Earth is there, when he suggested that gravity keeps the Moon in orbit around the Earth. The answer to Newton's problem is still not fully understood by scientists, but his formula can be used to great effect. It was of vital importance to NASA scientists when they made calculations in order to send the six Apollo Moon missions about fifty years ago.

Even before space travel, it was possible to use data about the Moon's **orbit** to work out the mass of the Earth. The time period of the Moon's orbit around the Earth is 27.3 days, or $T = 2.36 \times 10^6\,\text{s}$. The average orbital radius for the Moon is 384 000 km, or $r = 3.84 \times 10^8\,\text{m}$. From these data, we can calculate the mass of the Earth:

Gravitational attraction between Moon and Earth, $F = \dfrac{Gm_Em_M}{r^2}$

Centripetal force required to keep Moon in orbit, $F = \dfrac{m_Mv^2}{r}$

Gravity is the cause of the centripetal force, so these are equal:

$$\frac{Gm_Em_M}{r^2} = \frac{m_Mv^2}{r}$$
$$\frac{Gm_E}{r^2} = \frac{v^2}{r}$$
$$\frac{Gm_E}{r} = v^2$$
$$m_E = \frac{rv^2}{G}$$

EXAM HINT

Calculations about the properties of satellites usually need us to put the gravitational force expression equal to the centripetal force expression and then re-arrange.

$$\frac{Gm_1m_2}{r^2} = \frac{m_2v^2}{r}$$

The speed of the Moon comes from the time it takes to orbit:

$$v = \frac{2\pi r}{T}$$
$$= \frac{2\pi \times 3.84 \times 10^8}{2.36 \times 10^6}$$
$$v = 1022\,\text{m}\,\text{s}^{-1}$$

$$\therefore \quad m_E = \frac{3.84 \times 10^8 \times 1022^2}{6.67 \times 10^{-11}}$$
$$m_E = 6.01 \times 10^{24}\,\text{kg}$$

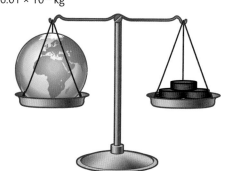

▲ **fig D** Weighing the Earth can be done by observing the Moon.

LEARNING TIP

Due to a wide variety of measuring devices and techniques, and rounding errors, you may see slight variations in the quoted mass of the Earth. They should all be close to $6 \times 10^{24}\,\text{kg}$.

CHECKPOINT

SKILLS ▶ INTERPRETATION

1. ▶ Explain why the weight of an object on the Earth is found by multiplying its mass in kilograms by 9.81.

2. Calculate the gravitational force between two π^0 particles in deep space, if they are 8 metres apart. The mass of a neutral pion is $2.40 \times 10^{-28}\,\text{kg}$.

3. Calculate the gravitational force between the Earth and the Moon if the Moon's mass is $7.35 \times 10^{22}\,\text{kg}$.

4. Calculate the average distance of the Earth from the Sun, if the mass of the Sun is $2.0 \times 10^{30}\,\text{kg}$.

5. ▶ Draw a sketch graph to show how the magnitude of the gravitational attraction, F, between two masses varies with the inverse square of the separation: $\dfrac{1}{r^2}$.

SUBJECT VOCABULARY

accrete to grow slowly by attracting and joining with many small pieces of rock and dust, due to the force of gravity

gravitational field a region of spacetime which is curved. This curvature will cause particles to experience an accelerating force

uniform acceleration acceleration that always has the same value; constant acceleration

orbit the curved path that a planet, satellite or similar object takes around another object in space

LEARNING OBJECTIVES

■ Derive and use the equation $g = \dfrac{Gm}{r^2}$ for the gravitational field strength due to a point mass.

■ Use the equation $V_{grav} = -\dfrac{GM}{r}$ for a radial gravitational field.

■ Compare electrical and gravitational fields.

RADIAL FIELDS

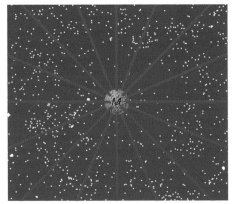

▲ **fig A** The radial gravitational field around a point mass.

Any mass will generate a gravitational field, which will then exert a force on any mass within the field. As gravity is always attractive, the field produced by a point mass will be radial in shape and the field lines will always point towards the mass (**fig A**).

GRAVITATIONAL FIELD STRENGTH

The radial field produced by a point mass naturally has its field lines closer together nearer the mass, as a result of its geometry (see **fig A**). This means that the strength of the field decreases with increasing distance from the mass causing it. The decrease is significant. In outer space, where it is the furthest possible from a galaxy or other particles, there are regions where there is almost no gravity. This can be explained mathematically by the formula which tells us the strength of a gravitational field at a certain distance, r, from a mass, M. We have already seen the force on a mass, m, caused by a gravitational field is $F = mg$. Also, the gravitational force on a mass, m, because of another mass, M, is given by Newton's expression:

$$F = \frac{GMm}{r^2}$$

These two expressions are calculating the same force, so must themselves be equal:

$$F = \frac{GMm}{r^2} = mg$$

$$\therefore \quad g = \frac{GM}{r^2}$$

i.e. the field strength is independent of the object being acted upon.

WORKED EXAMPLE 1

Calculate the gravitational field strength caused by the Earth at the distance of a geostationary satellite, which orbits at a height, h, of 35 800 km above the surface of the Earth. The radius of the Earth, R_E, is 6400 km. From **Section 11A.1** we have the mass of the Earth as 6.01×10^{24} kg.

Distance from the Earth's centre:

$$r = R_E + h = 42\,200 \text{ km} = 4.22 \times 10^7 \text{ m}$$

$$g = \frac{GM}{r^2}$$

$$= \frac{6.67 \times 10^{-11} \times 6.01 \times 10^{24}}{(4.22 \times 10^7)^2}$$

$$\therefore \quad g = 0.23 \text{ N kg}^{-1}$$

GRAVITATIONAL POTENTIAL

Potential energy is the stored energy that an object has due to its position. The potential at a point in any type of field is the amount of energy which is needed to get to that position in the field for any object which is affected by the field. Therefore for a gravitational field it is expressed as an amount of energy per unit mass (J kg^{-1}), as mass is what is affected by the field. The definition of **gravitational potential** is the amount of work done per unit mass to move an object from an infinite distance to that point in the field. As gravitational fields are always attractive, objects will always gain energy on moving into a point in the field, and so gravitational potential is always a negative quantity. Gravitational potential can be calculated at a distance r from a mass M from the equation:

$$V_{grav} = -\frac{GM}{r}$$

WORKED EXAMPLE 2

Calculate the gravitational potential caused by the Earth, at the distance of a geostationary satellite, which orbits at a height, h, of 35 800 km above the surface of the Earth. The radius of the Earth, R_E, is 6400 km.

$$r = R_E + h = 42\,200 \text{ km} = 4.22 \times 10^7 \text{ m}$$

$$V_{grav} = -\frac{GM}{r}$$

$$= \frac{-6.67 \times 10^{-11} \times 6.02 \times 10^{24}}{(4.22 \times 10^7)}$$

$$V_{grav} = -9.52 \text{ MJ kg}^{-1}$$

The gravitational field strength tells us how quickly the potential is changing over distance, and the mathematical connection between the two is:

$$g = -\frac{dV_{grav}}{dx}$$

For small distances, in which g does not change significantly, this can be calculated using the actual distance change:

$$g = -\frac{\Delta V_{grav}}{\Delta x}$$

$$g\Delta x = -\Delta V_{grav}$$

As the potential was defined per unit mass, and is already a negative value, the magnitude of the actual gravitational potential energy change for an object of mass m will then be given by:

$$E = m \times \Delta V_{grav} = mg\Delta x$$

which we have previously used (see **Book 1, Section 1B.1**) as:

$$\Delta E_p = mg\Delta h$$

ELECTRIC AND GRAVITATIONAL FIELDS

You may have noticed something familiar about the equations we have been using to calculate gravitational forces and fields. Mathematically, they are of exactly the same form as those we used to calculate the force between charged particles, electric potential and electric field strength. The only differences are the symbols we use to represent the quantities causing the fields, and the constants of proportionality. The similarities come from the fact that both types of field are radial from a point. The constants of proportionality depend upon the way the forces interact with the Universe, and with the unit system that we use in the calculations.

QUANTITY	GRAVITATIONAL FIELD	ELECTRICAL FIELD
force	$F = \dfrac{Gm_1m_2}{r^2}$	$F = \dfrac{kq_1q_2}{r^2}$
field strength	$g = \dfrac{Gm}{r^2}$	$E = \dfrac{kq}{r^2}$
potential	$V_{grav} = -\dfrac{Gm}{r}$	$V = \dfrac{kq}{r}$

table A Gravitational and electric fields compared. Remember, the Coulomb law constant $k = \dfrac{1}{4\pi\varepsilon_0} = 9.0 \times 10^9 \, N \, m^2 \, C^{-2}$.

Although the force and field strength of the gravitational and electrical fields share the same form, they do differ in some significant ways. Gravitational forces are always attractive but electric forces are not. As electrical charges can be either positive or negative, the electric field can be in either direction to or from a charge. A charged particle can be shielded from an electric field, but a massive particle cannot be shielded from a gravitational field. In addition, the electromagnetic force is significantly stronger than the gravitational force.

DID YOU KNOW?

Gravity is the weakest of the four fundamental forces in nature (see **Section 7C**), but it is the only one that acts on all matter particles, and at all distances.

WORKED EXAMPLE 3

An electron and a proton experience electrical attraction and gravitational attraction. When the electron is 10 metres from the proton the electrostatic attraction is F. Calculate the distance of the electron from the proton when the gravitational attraction is also equal to F. Comment on the answer.

Electrostatic force:

$$F = \frac{kq_1q_2}{r^2}$$

$$= \frac{9.0 \times 10^9 \times 1.6 \times 10^{-19} \times -1.6 \times 10^{-19}}{10^2}$$

$$\therefore F = -2.3 \times 10^{-30} \, N$$

Gravitational attraction:

$$F = \frac{GMm}{r^2}$$

$$= 2.3 \times 10^{-30} \, N$$

$$r^2 = \frac{Gm_pm_e}{2.3 \times 10^{-30}}$$

$$= 4.41 \times 10^{-38}$$

$$r = 2.1 \times 10^{-19} \, m$$

The distance between the two centres of gravity, therefore, is smaller than the radius of a proton, so it would be impossible for the electron to get this close. Gravity is an extremely weak force in comparison to the other three forces in nature, but it affects all matter, not just things that are charged.

CHECKPOINT

1. Calculate the gravitational field strength and gravitational potential at the orbit of a polar satellite that travels at 900 km above the surface of the Earth.

    SKILLS PROBLEM SOLVING, ANALYSIS

2. Compare the magnitudes of:
    (a) the force of attraction between an electron and a proton in a hydrogen atom (radius = 0.53×10^{-10} m), with
    (b) the gravitational attraction of the Sun on Pluto. Pluto has a mass of 1.29×10^{22} kg and orbits at an average distance of 5900 million kilometres from the Sun.

    SKILLS CRITICAL THINKING

3. If the Earth's gravity exerts a force of attraction on the Moon equal to the electrostatic attraction between the proton and the electron in a hydrogen atom, then what would the mass of the Moon be?

    SKILLS PROBLEM SOLVING, CRITICAL THINKING

4. There is a point on the line between the centres of the Earth and the Moon where their gravitational fields have equal magnitude but are in opposite directions, effectively creating a point of zero gravity. Calculate the distance of this point from the centre of the Earth. (Mass of the Moon = 7.35×10^{22} kg)

SUBJECT VOCABULARY

gravitational potential the amount of work done per unit mass to move an object from an infinite distance to that point in the field

MONITORING VOLCANOS

SKILLS CRITICAL THINKING, PROBLEM SOLVING, REASONING/ ARGUMENTATION, INTERPRETATION, ADAPTIVE LEARNING, ADAPTABILITY, CONTINUOUS LEARNING, PRODUCTIVITY, COMMUNICATION

Volcanology is the study of volcanoes. The importance of detailed scientific research into the development and explosions of volcanoes was highlighted by the 2010 eruption of Eyjafjallajökull in Iceland, which closed much of Europe's airspace for nearly a week.

The extract below is a section from the book *Introducing Volcanology*. It explains how internal changes within a volcano can be monitored by their effects on the local gravitational field strength. These internal changes might be early signs that an eruption will occur.

POPULAR SCIENCE BOOK

GRAVITY AND ELECTROMAGNETIC MONITORING

The overall structure and even slight changes in the character of a volcano at more local scales can be studied using gravity. A general survey of the gross volcanic structure using gravity can provide an understanding of any underlying anomalies that may relate to the presence of dense or light bodies beneath the structure. Gravity measurements may also be used to get much closer to understanding what's happening in the magma plumbing before and even during eruptions. To do this a technique known as **microgravity** provides much more necessary detail. This is a geophysical technique that can measure minute changes in the Earth's gravity. It must be corrected to take account of any ground movements that are occurring at the same time, as these will have an effect on the overall gravity measurement. Filling of a magma chamber with new magma, or the **vesiculation** of an existing magma reservoir, prior to eruption, will change the density of the surroundings and can be picked up with microgravity. **Electromagnetic fields (EM)** are also used to provide quite a sensitive source of measurement in the shallow parts of a volcanic system, partly due to limitations on the penetration depth that can be recorded. EM signals can be affected by the **electrical resistivity** which can be changed by fluids, and changes in the geomagnetic field, altered by magma movement and magma type. These measurements, when combined with other remote sensing techniques, can be very powerful tools with which to monitor the shallow volcanic system.

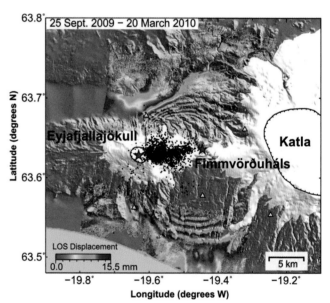

▲ **fig A** Mapping volcanic land deformation around the Fimmvörðuháls volcano in Iceland, using interference patterns (the coloured bands) between satellite maps from different times.

From pages 86–87 and page 89 (in Chapter 9) of *Introducing Volcanology* by Dr Dougal Jerram, ISBN 978-1-90671-622-6, published by Dunedin Academic Press (2011).

SCIENCE COMMUNICATION

The preface of this book says that it is aimed at 'those with an inquisitive interest as well as the more advanced reader'.

1 Explain how the author has written at a scientific level appropriate to his target audience. What has he assumed about the reader's capability, and in which areas has he restricted the depth of explanation?

2 Why do you think some words and phrases (for example 'vesiculation') are in bold print in the text?

3 The image in **fig A** is from two pages after the text, and accompanies a different section in the book. What is its relevance to the text quoted here?

PHYSICS IN DETAIL

Now we will look at the physics in detail. Some of these questions will link to topics elsewhere in this book, so you may need to combine concepts from different areas of physics to work out the answers.

4 We say that the gravitational field strength at the Earth's surface is $9.81 \, \text{N kg}^{-1}$.

(a) Write the equation for the gravitational field strength at the surface of a planet.

(b) What assumptions are made in using this equation, which are not really accurate for planet Earth?

(c) Which of these assumptions would be affected by the movements of magma and gases within a volcano, which cause the variation of surface gravity, as described in the extract?

(d) In terms of the equation, how could ground movements also affect the gravitational field strength?

5 (a) Explain how the colour bands in **fig A** illustrate changes in ground level over time.

(b) Why would any variations in ground level not affect the orbital radius of the satellite that took measurements to generate **fig A**?

(c) What scientific strength is there in comparing the conclusions from **fig A** and those from a microgravity survey as described in the text?

6 The section about electromagnetic field testing suggests that the EM signals will be differently affected by areas of differing resistivity.

(a) Explain how the electrical resistivity of regions within a volcano might change over time.

(b) Explain how variations in resistivity could alter the penetration by electric fields.

ACTIVITY

Write a design brief for an experiment to measure variations in gravitational field strength. Your experiment should be capable of monitoring over time, and also over a range of geographical locations around a volcano. Include instructions about the measurements the equipment needs to make, and what is required to be able to make conclusions.

1 Which statement is correct?

 A Gravitational field strength is a scalar quantity.

 B Electrostatic field strength is a scalar quantity.

 C Gravitational and electrostatic forces both follow an inverse square law.

 D Gravitational and electrostatic forces are both contact forces.   [1]

 (Total for Question 1 = 1 mark)

2 Which is a reasonable estimate of the gravitational force of attraction between you and your chair?

 A 4×10^{-7} N

 B 80 kg

 C 785 N

 D 6400 N   [1]

 (Total for Question 2 = 1 mark)

3 The Orion spacecraft consists of a capsule (mass = 8913 kg) and a service module (mass = 12 337 kg) and is designed to carry four astronauts (mass 80.0 kg each). Calculate the minimum energy needed to move the fully manned spacecraft completely outside the influence of the gravitational field of the Earth.

 Mass of the Earth = 6.00×10^{24} kg
 Radius of the Earth = 6.40×10^{6} m
 $G = 6.67 \times 10^{-11}$ N m^2 kg^{-2}

 A 2.12×10^{3} J

 B 6.30×10^{6} J

 C 1.34×10^{12} J

 D 1.36×10^{12} J   [1]

 (Total for Question 3 = 1 mark)

4 Which of the following expressions correctly gives the distance, R, from a mass, M, where its gravitational potential will equal V?

 A $R = -GMV_{grav}$

 B $R = -\dfrac{GM}{V_{grav}}$

 C $R = -\dfrac{M}{GV_{grav}}$

 D $R = -\dfrac{G}{MV_{grav}}$ [1]

 (Total for Question 4 = 1 mark)

5 In a physics lesson a student learns that the Earth is 81 times more massive than the Moon. Searching the Internet, he is surprised to discover that the gravitational field strength at the surface of the Earth is only 6 times greater than that at the surface of the Moon.

 Use the above data to compare the radius of the Earth with that of the Moon.   [3]

 (Total for Question 4 = 3 marks)

6 (a) Derive an expression for the gravitational field strength g at a distance r from the centre of mass M.   [2]

 (b) Use your expression to calculate g at the surface of the Earth.

  mass of Earth $M_E = 5.97 \times 10^{24}$ kg
  radius of Earth $r_E = 6.38 \times 10^{6}$ m   [1]

 (Total for Question 5 = 3 marks)

7 The Hubble Space Telescope (HST) was launched in 1990 into an orbit of radius 6940 km. The satellite makes 15 complete orbits of the Earth every 24 hours and its position high above the Earth's atmosphere has allowed high quality images of extremely distant objects to be produced.

 (a) Show that the HST has a centripetal acceleration of about 8 m s^{-2} [4]

 (b) The HST is kept in orbit by the gravitational pull of the Earth. Use your answer to (a) to calculate a value for the mass of the Earth.   [3]

 (Total for Question 7 = 7 marks)

8 Communications satellites were first proposed in 1945 by the science fiction author Arthur C. Clarke. In an article published in the magazine *Wireless World* he asked whether rocket stations could give worldwide radio coverage.

In the article Clarke states:

'There are an infinite number of possible stable orbits, circular and elliptical, in which a rocket would remain if the initial conditions were correct. A velocity of 8 km s⁻¹ applies only to the closest possible orbit, one just outside the atmosphere, and the period of revolution would be about 90 minutes. As the radius of the orbit increases the velocity decreases, since gravity is diminishing and less centrifugal force is needed to balance it.'

With permission of *Electronics World* www.electronicsworld.co.uk

(a) State what is meant in the article by the phrase 'gravity is diminishing', and criticise the statement that 'less centrifugal force is needed to balance (the satellite)'. [3]

(b) (i) By deriving an appropriate equation, show that the orbital speed of the satellite decreases as the radius of orbit increases. [3]

   (ii) By deriving an appropriate equation, show that the orbital period of a satellite increases as the orbital speed decreases. [2]

(c) The period T of a satellite in a circular orbit is given by the equation:

$$T = \sqrt{\frac{4\pi^2 r^3}{GM}}$$

where r is the radius of orbit and M is the mass of the Earth.

Calculate the period of a satellite in an orbit 4.0×10^5 m above the surface of the Earth.

mass of the Earth = 5.98×10^{24} kg

radius of the Earth = 6.36×10^6 m [2]

(d) After a time the radius of the satellite's orbit will start to decrease due to the resistive forces acting on the satellite from the atmosphere. As this happens the satellite speeds up.

Describe the energy changes occurring as the radius of the orbit decreases. [2]

(Total for Question 8 = 12 marks)

9 (a) State what is meant by the term gravitational field and write an equation for the gravitational strength acting on a mass in this room. [2]

(b) Sketch a graph of the gravitational potential, V_{grav} at increasing distance, r, from a mass, M. [2]

(c) Calculate the gravitational potentials at the surface of the Earth due to:

   (i) the Sun [1]

   (ii) the Earth. [1]

   Mass of the Earth = 6.0×10^{24} kg
   Mass of the Sun = 2.0×10^{30} kg
   Radius of the Earth = 6.4×10^6 m
   Radius of the Earth's orbit = 1.5×10^{11} m
   G = 6.67×10^{-11} N m² kg⁻²

   (Total for Question 9 = 6 marks)

10 Explain the similarities and differences between gravitational and electric fields. [6]

(Total for Question 10 = 6 marks)

TOPIC 11 ASTROPHYSICS AND COSMOLOGY

11B SPACE

Humans have always wondered at the stars in the night sky. Stories from long ago about constellations, paintings and recent photographs, have all added to the mystery.

Scientists have applied the usual techniques of measurement and evidence-seeking to the studies of astronomy and astrophysics. It is remarkable how much information we have gained about the stars, galaxies and the Universe as a whole, from only the light that lands on us from these distant celestial bodies. We can use this information to infer the existence of many things we cannot see. Most of the Universe is dark energy and dark matter, and although we cannot detect either of these, starlight can tell us they are there and affecting everything in the Universe.

In this chapter, we will look at how stars emit energy as electromagnetic radiation of various wavelengths, and what we can find out from the measurements of this radiation when it arrives at the Earth. We will see how to determine the distance to a star, how hot it is, what type of star it is, what it is made of, and also much more.

MATHS SKILLS FOR THIS CHAPTER

- **Use of angles in 2D and 3D structures** (*e.g. using trigonometric parallax*)

- **Use of small angle approximations** (*e.g. in trigonometric parallax*)

- **Solving algebraic equations** (*e.g. finding the temperature of a star from its peak wavelength*)

- **Changing the subject of a non-linear equation** (*e.g. finding the temperature of a star from its luminosity*)

- **Sketching and interpreting relationships shown graphically** (*e.g. comparing the Stefan–Boltzmann curves for stars of different temperatures*)

What prior knowledge do I need?

Topic 1A (Book 1: IAS)
* Newton's laws of motion

Topic 1B (Book 1: IAS)
* Conservation of energy

Topic 3D (Book 1: IAS)
* Emission and absorption spectra

Topic 8A
* Heat and temperature

Topic 11A
* Gravitational forces

What will I study in this chapter?

* The life cycles of stars
* How to calculate the energy emitted by stars
* Stellar classification
* How to measure the distances to stars and galaxies
* Redshift and Hubble's law
* The development of galaxies and the Universe as a whole
* Dark matter and dark energy

What will I study later?

Study beyond International A Level
* The development of planets and star systems
* The interactions between galaxies
* The mapping of dark matter
* Experimental research into dark matter
* The structure and operation of telescopes

STELLAR PROPERTIES

Stars are at such large distances from the Earth that the only information we have about them is the electromagnetic radiation we receive from them (**fig A**). However, from this limited information, we can measure various stellar properties. These allow us to classify stars into various groups which have quite strange names, such as **red giant**, **white dwarf** and **blue supergiant**. They are much too far away for us to send probes to them, or even to send signals to them in the hope of detecting reflections. However, scientists have managed to determine a large amount of detailed information from even the faintest glows in the night sky. Incredibly, the electromagnetic emissions from stars can tell us their temperature, chemical composition, speed of movement, approximate age, size and much more.

▲ **fig A** Starlight is the only information we have about the stars.

BLACK BODY RADIATION

A **black body radiator** is a theoretical body which completely absorbs all radiation that lands on it. We can imagine a cavity with a tiny hole in one side, and the hole itself is the black body. It is considered to absorb all radiation that lands on the hole, as it goes into the box (or cavity) and bounces around with no chance of escaping out of the hole before it is absorbed inside the box. No radiation is reflected, hence the name 'black body'.

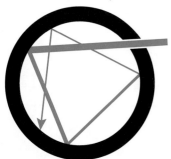

▲ **fig B** A box with a tiny hole in the surface is considered to be a perfect black body.

All objects emit electromagnetic radiation. Infrared cameras can detect radiation from our bodies, but with our eyes we can only see visible EM radiation from hot objects; for example, the red glow from a fire. We know that good absorbers of radiation are also good emitters, and a black body radiator is considered to be a perfect emitter. It will emit EM radiation at all wavelengths, and how much of each wavelength will depend on the black body's temperature.

THE STEFAN–BOLTZMANN LAW

How bright is a star? With the naked eye, we are only able to distinguish six different levels of how bright stars appear to us (**fig C**). This is insufficient for scientific use, as many stars of differing brightness would appear identical to our eyes. Astronomers therefore use a more precise measure to classify the actual brightness of stars: their output power, which is known as **luminosity**.

▲ **fig C** Can you identify the brightest and dimmest stars in this star cluster, NGC 290?

We define luminosity as the rate at which energy of all types is radiated by an object in all directions. This depends upon both the object's size and, more importantly, its temperature.

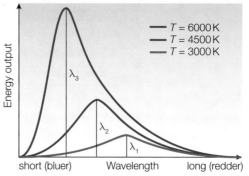

▲ **fig D** Black body radiation curves for different temperatures.

The electromagnetic radiation is emitted across a very large range of wavelengths. A perfect black body radiator will emit energy across the entire electromagnetic spectrum, following a distribution like those in **fig D**. This distribution is given by the **Stefan–Boltzmann law**. This tells us that the output power from a black body is proportional to its surface area and the fourth power of its temperature in kelvin.

$$L = \sigma A T^4$$

where the Stefan–Boltzmann constant, $\sigma = 5.67 \times 10^{-8}$ W m^{-2} K^{-4}.

For a sphere, this would become:

$$L = 4\pi r^2 \sigma T^4$$

Working on the assumption that a star acts like a black body emitter, which is a very good approximation, this equation describes the luminosity of a star.

LEARNING TIP

A black body radiator is a theoretical perfect emitter, which follows the Stefan–Boltzmann energy output curve for its temperature, as illustrated in **fig C**, and also follows Wien's law below. Here we are approximating stars to perfect black bodies. Remember, black body radiation is a thermodynamics idea that can be applied in other areas of physics.

WORKED EXAMPLE 1

The Sun has a radius of 7.0×10^8 m and its surface temperature is 5800 K. What is the luminosity of the Sun?

Luminosity of the Sun, L:

$$L = \sigma A T^4$$
$$= \sigma 4\pi r^2 T^4$$
$$= 5.67 \times 10^{-8} \times 4\pi \ (7.0 \times 10^8)^2 \times 5800^4$$
$$= 3.95 \times 10^{26} \, \text{W}$$

What would be the surface temperature of a star of radius 14.0×10^8 m with the same luminosity as the Sun?

$$A_{star} = 4\pi r^2$$
$$= 4\pi \times (14.0 \times 10^8)^2$$
$$= 2.46 \times 10^{19} \, \text{m}^2$$

$$T_{star} = \sqrt[4]{\frac{L}{\sigma A}}$$
$$= \sqrt[4]{\frac{3.95 \times 10^{26}}{(5.67 \times 10^{-8} \times 2.46 \times 10^{19})}}$$
$$= 4100 \, \text{K}$$

Therefore a bigger star can be at a lower temperature and yet have the same luminosity, i.e. it looks as bright.

WIEN'S LAW

To calculate luminosity, we needed to know the temperature of the star. There are various methods for determining the temperatures of stars, but we will focus on one that uses the wavelengths of light given off by a star. When we examine the range of wavelengths emitted by a star, known as its spectrum, we find that some wavelengths are given off with more intensity than others.

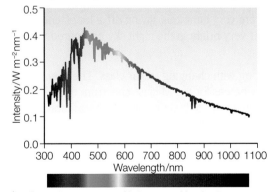

▲ **fig E** The spectrum of light emitted by the Sun.

We saw from the Stefan–Boltzmann law that as the temperature of a black body increases, it emits more energy. **Fig D** shows how the amount of energy emitted at different wavelengths changes with temperature. At higher temperatures the curve has a more pronounced peak, and the wavelength of the peak output gets

shorter as the temperature rises. The relationship between the peak output wavelength and temperature is described by **Wien's law**:

$$\lambda_{max} T = 2.898 \times 10^{-3} \, \text{m K}$$

The number 2.898×10^{-3} m K is known as Wien's constant.

WORKED EXAMPLE 2

Looking at the spectrum of light from Betelgeuse, in the constellation of Orion, its peak wavelength is at 9.35×10^{-7} m. What is the surface temperature of Betelgeuse?

$$\lambda_{max} T = 2.898 \times 10^{-3}$$
$$T = \frac{2.898 \times 10^{-3}}{\lambda_{max}}$$
$$= \frac{2.898 \times 10^{-3}}{9.35 \times 10^{-7}}$$
$$T = 3100 \, \text{K}$$

CHECKPOINT

1. What is the luminosity of the star Sirius, which has a surface temperature of 12 000 K and a diameter of 2 220 000 km?

2. (a) The Sun's surface temperature has been measured as being 5800 K. Calculate the peak wavelength of the solar spectrum.
   (b) The peak wavelength from the Sun is in fact measured at ground level as being 470 nm. Why is the measured value different from your calculated value from part (a)?

   SKILLS   ANALYSIS

3. ▶ Calculate the peak wavelength output we would expect of Bellatrix if its surface temperature is 21 500 K.

   SKILLS   INTERPRETATION

4. ▶ From **fig F**, calculate the temperatures of Canopus and Rigel.

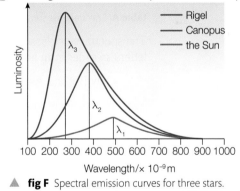

▲ **fig F** Spectral emission curves for three stars.

SUBJECT VOCABULARY

red giant a large star which is cooler than our Sun, e.g. 3000 K.
white dwarf a small, hot star, perhaps 10 000 K
blue supergiant a very large, very hot star, perhaps 25 000 K
black body radiator a theoretical object, that completely absorbs all radiation that lands on it
luminosity the rate at which energy of all types is radiated by an object in all directions
Stefan–Boltzmann law the power output from a black body is proportional to its surface area and the fourth power of its temperature in kelvin, $L = \sigma A T^4$
Wien's law the relationship between the peak output wavelength and temperature for a black body radiator is given by the equation: $\lambda_{max} T = 2.898 \times 10^{-3}$ m K

LEARNING OBJECTIVES

■ Sketch and interpret a simple Hertzsprung-Russell diagram relating stellar luminosity to surface temperature.

■ Relate the Hertzsprung-Russell diagram to the life cycle of stars.

STAR CLASSES

Astronomers have classified stars into groups according to their temperature. This is a useful property to use since stars with similar temperatures tend to share many other features. As we saw from Wien's law (see **fig E** in **Section 11B.1**) the temperature determines the spectral output of the star, but it can also suggest chemical composition and age.

SPECTRAL CLASS	EFFECTIVE TEMPERATURE/K	COLOUR	M/M_{SUN}	R/R_{SUN}	L/L_{SUN}	MAIN SEQUENCE LIFESPAN
O	28 000–50 000	blue	20–60	9–15	90 000–800 000	1–10 Myr
B	10 000–28 000	blue-white	3–18	3.0–8.4	95–52 000	11–400 Myr
A	7500–10 000	white	2.0–3.0	1.7–2.7	8–55	400 Myr–3 Gyr
F	6000–7500	white-yellow	1.1–1.6	1.2–1.6	2.0–6.5	3–7 Gyr
G	4900–6000	yellow	0.85–1.1	0.85–1.1	0.66–1.5	7–15 Gyr
K	3500–4900	orange	0.65–0.85	0.65–0.85	0.10–0.42	17 Gyr
M	2000–3500	red	0.08–0.05	0.17–0.63	0.001–0.08	56 Gyr

table A Spectral class summary.

The spectral star classes are labelled by letters (**table A**). One mnemonic for remembering the letters, in order from hottest to coolest, is:

> Only Brilliant Astronomers Fix Grave Kepler Mistakes

You will notice a number of trends in the data in **table A**. The hotter stars often have more mass and are more luminous. The larger gravitational pressure at the centre of a massive star makes the nuclear fusion reactions within the star run very fast, producing a lot of energy and using the hydrogen fuel in the star at a very high rate. In addition, more massive stars are also larger. Therefore, with large size and high temperatures, the hotter stars are very luminous, giving off a lot of energy. At the same distance away from us, these will then appear very bright in the night sky compared with a smaller, cooler star.

Table A also shows the change in colour that goes along with changing stellar class. The overall impression of the O class spectrum is a bluish colour, whereas for the M class spectrum there is a strong red colour showing (also see **fig A**). These changes are a direct result of the variation in output curve we saw associated with Wien's law. **Fig B** highlights how this comes about.

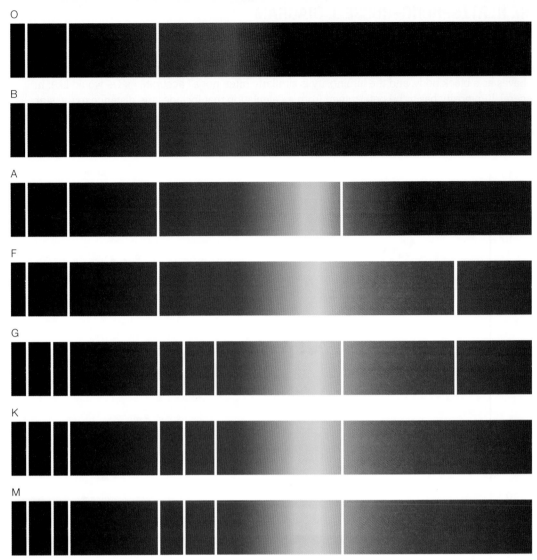

▲ **fig A** Light spectra from typical stars in each star class. Note the white lines, which highlight the position of absorption lines (actually black in the original photos). These give us information about the chemical composition of the stars, and the Doppler shift which tells us their velocity.

Looking at **fig B**, if we combine together equal intensity parts of the colours of the visible spectrum, this will produce white light. The extra emissions above this white basic emission – the peak of the curve on the graph – will show the colour the star appears. Also, given the pattern of decreasing size with luminosity we have seen above, it is logical to find that blue supergiants are generally O class stars, whereas red dwarf stars are usually in the M class.

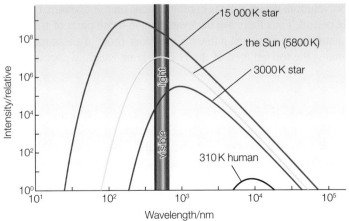

▲ **fig B** The peak of a star's Wien's law curve shows us what colour it will appear.

THE HERTZSPRUNG–RUSSELL DIAGRAM

If you plot a graph of luminosity against star temperature to confirm the trend of **table A**, you will find that there is a general correlation. However, the more data you add, the more complex the picture becomes. We must also remember that the temperature measurement assumes the star behaves as a black body, and the luminosity is similarly often not as accurate as we would like. In general, luminosity has to be determined from a calculation that includes the distance to the star, and measuring the distances to stars is by no means an exact science (see **Section 11B.3** for more details). Such a plot can give us some very useful insights, but is not a graph in the true sense. It is known as a **Hertzsprung–Russell diagram** (**fig C**).

EXAM HINT

You will be expected to be able to sketch a simple Hertzsprung–Russell diagram from memory in the exam. This should include approximate values for the axes.

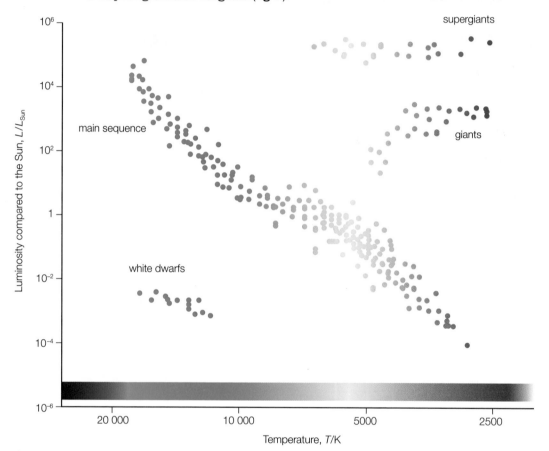

▲ **fig C** A Hertzsprung–Russell diagram.

Most stars we observe fall on a diagonal line across the Hertzsprung–Russell diagram, which is called the **main sequence**. These **main sequence stars** are stable stars which will exist in this state for the majority of their lifetime. Their correlation represents the connection between brightness and high temperature. Note that the plot is always drawn with hotter temperatures on the left-hand side. There are also other stages in a star's evolution, which appear in other places on the diagram, but these are much shorter than its stable period. Thus there are far fewer stars in those parts of the H–R diagram. For example, as blue supergiants burn out in just a few million years, most of these have already done so, and we do not see many; the top left area of the H–R diagram is lightly populated.

STELLAR EVOLUTION

The majority of ordinary matter in the Universe is hydrogen ($\approx$ 75%) or helium ($\approx$ 24%), and it is from these elements that stars are initially formed. From an accreting collection of these gases, called a **protostar**, the life cycle of a star follows a number of stages, with the star ending its life as a white dwarf, **neutron star** or **black hole**. As the star undergoes nuclear fusion, the binding energy differences of the nuclei before and after fusion mean that the process releases energy, often as electromagnetic radiation, to heat the star. The structure of the star is protected from gravitational collapse by the pressure from the vibration of its particles and the electromagnetic radiation which is trying to escape. It is this constant battle between the outward pressure and gravity that drives the evolution of a star throughout its lifetime. The initial mass of the star is a critical factor in determining how the battle goes, and which of the possible life cycles a star will follow (**fig D**).

The multiple possible life cycles for stars are usually grouped together into just two paths in which the outcomes are similar. These are the life cycle for low-mass stars (such as our Sun) and the life cycle for massive stars, which have at least four times the mass of our Sun.

LOW-MASS STARS

Once it has accreted about the mass of our Sun, a low-mass star will start to undergo nuclear fusion of hydrogen, converting this into helium. This is a stable stage of life in which radiation pressure and gravity are in equilibrium. The star will remain in this state for billions of years. Eventually, it will run low on hydrogen fuel, but will have produced so much energy that it will expand slightly. This expansion causes the temperature to fall and the star becomes a red giant. Our Sun is expected to undergo this change in another 4–5 billion years, when it will expand beyond the orbit of Venus. Once most of the hydrogen fuel is used, the star will start fusing helium nuclei. This complex process can cause an explosion which throws some material from the star out into space, forming a **planetary nebula**. As the fuel to produce energy to support the star runs out, the outward pressure from fusion drops and gravity takes hold, causing the star to contract to a much smaller size. This heats up the star significantly and it becomes a white dwarf. As time continues, the star will slowly run out of energy and die, passing through the red dwarf stage to become a **black dwarf**. Note that the black dwarf stage is theoretical, as it takes a white dwarf longer than the current age of the Universe to cool this much, so there has not yet been time for any to develop.

MASSIVE STARS

If a protostar has more than four times the mass of our Sun, the star begins life as a blue supergiant. As with low-mass stars, nuclear fusion begins and the star enters a stable stage of life in which heat pressure and gravity are in equilibrium. However, the fusion processes happen at much higher temperatures than in lower mass stars. The core of our Sun may be at 15 million kelvin, whereas a large star could have a core temperature of 40 million kelvin. This means that it burns very quickly, and the conditions make it possible for further fusion of some of the larger nuclei it produces to occur. The fusion of helium can produce a variety of the larger elements, which have mass numbers which are multiples of 4 (helium has four nucleons), such as carbon, oxygen and silicon. There will then be stages of carbon and oxygen fusing. A high-mass star is likely to be on the main sequence for only up to a billion years. When the material of such a star has been fused to become mostly iron, it can no longer undergo nuclear fusion and it stops producing energy. This happens even more suddenly than in a low-mass star, and with the enormous gravitational forces produced by the large mass, it undergoes a major collapse. This sudden increase in density produces a sudden huge burst of energy, effectively bouncing the collapse back out. This explosion is called a (Type II) **supernova** and is the most immense burst of energy ever witnessed. It is so bright, at about 10 billion times the luminosity of the Sun, that you can see the change in the night sky with the naked eye.

Within a supernova explosion there is so much energy that nuclear reactions occur that produce the elements above iron in the periodic table. The natural occurrence of these elements is evidence that supernovae must have occurred in the past, as the binding energies of these heavy elements are such that they cannot be created in other natural processes in the Universe.

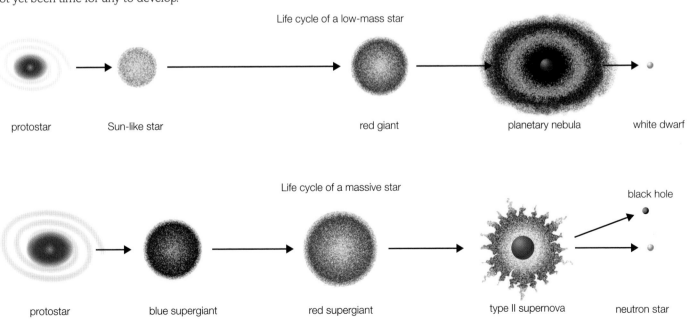

Life cycle of a low-mass star

protostar   Sun-like star   red giant   planetary nebula   white dwarf

Life cycle of a massive star

protostar   blue supergiant   red supergiant   type II supernova   black hole   neutron star

▲ **fig D** How initial mass affects stellar evolution.

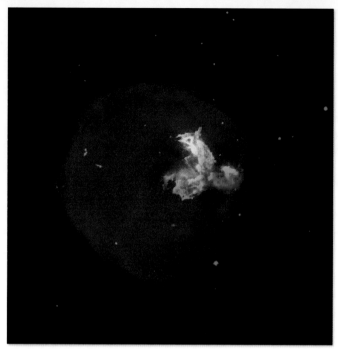

▲ **fig E** The remains of supernova N63A Menagerie.

After a high-mass star has exploded as a supernova, the entire star may be completely shattered. If there remains a central core of stellar material, this will either be a neutron star (if the core was up to three solar masses) or a black hole (if the core kept more than three solar masses). Neither of these is easy to detect, as they emit little or no light, and they are not plotted on the Hertzsprung–Russell diagram. A neutron star consists almost entirely of neutrons, packed as densely together as the nucleons within the nucleus of an atom. They can hold three times the mass of the Sun but are only about 10 km in diameter. Black holes are even smaller and hold even more matter than neutron stars. This means that their gravitational pull is immense, so strong that even things travelling at the speed of light cannot escape.

You will see from **fig F** that a given star develops through its life as various different types of star. If we observe it at each of these points in its life, they would be plotted in different places on the H–R diagram. Thus, we can plot the life cycle of a star as a movement around the H–R diagram (**fig F**).

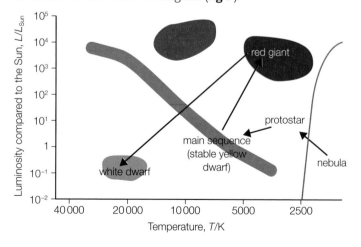

▲ **fig F** The Sun's life cycle will move it around the H–R diagram.

SUBJECT VOCABULARY

Hertzsprung–Russell diagram a plot of stars, showing luminosity (or absolute magnitude) on the *y*-axis, and temperature (or spectral class) on the *x*-axis

main sequence a diagonal line from top left to bottom right of a Hertzsprung–Russell diagram which marks stars that are in a generally stable phase of their existence

main sequence star a stable star whose core performs hydrogen fusion and produces mostly helium

protostar the mass of dust and gas clumping together under the force of gravity prior to the start of nuclear fusion in its core, which will go on to become a star

neutron star one of the possible conclusions to the life of a large mass star; small and very dense, composed of neutrons

black hole one of the possible conclusions to the life of a large mass star; a region of space–time in which the gravity is so strong that it prevents anything from escaping, including EM radiation

planetary nebula the remains of an explosion at the end of the life cycle of a low-mass star; material which may eventually join together into new planets

black dwarf the final stage of the life cycle of a small mass star, when nuclear fusion has ceased and it has cooled so that it no longer emits visible light

supernova the explosion of a large mass star at the end of its lifetime, when it becomes extremely unstable

LEARNING OBJECTIVES

■ Determine astronomical distances using trigonometric parallax.

■ Use the equation for the intensity of a star, $I = \dfrac{L}{4\pi d^2}$.

■ Measure astronomical distances using *standard candles*.

BIG DISTANCE UNITS

Astronomical distances are very large. Our nearest neighbour is the Moon, and even that is nearly 400 000 000 m away. The Sun is 150 000 000 000 m away, and the distance to the orbit of Neptune is 4 500 000 000 000 m. Measuring across space generates very large values. Using standard form notation helps with this, but astronomers have defined a number of alternative distance units to cut down the magnitudes of the numbers involved.

You have probably already heard of the **light year**. This is the distance that light can travel in one year, which is about 10^{16} m. We also use the **astronomical unit (AU)**, which is the radius of the Earth's orbit around the Sun: 1 AU = 1.5×10^{11} m.

Much of our understanding of the structure and formation of galaxies depends on being able to measure the distances to stars accurately. Astronomers have developed a number of techniques for doing this, but all have their limitations. These limitations can be overcome, or at least minimised, by comparing the results from the different techniques on the same star, and refining the techniques to improve accuracy.

LEARNING TIP

Some of the techniques explained in this section are used on individual stars, and some are used on large groups of stars, such as galaxies or star clusters.

TRIGONOMETRIC PARALLAX

To measure the distance to relatively close stars, astronomers use a method which is commonly used in surveying, known as **trigonometric parallax**. As the Earth moves around the Sun, a relatively close star will appear to move across the background of more distant stars. This optical illusion is used to determine the distance of the star. The star itself does not move significantly during the course of the observations. To determine the trigonometric parallax you measure the angle to a star, and observe how that changes as the position of the Earth changes. We know that in six months the Earth will be exactly on the opposite side of its orbit, and therefore will be two astronomical units from its location today.

Using observations of the star which is to be measured against a background of much more distant stars, we can measure the angle between the star and the Earth in these two different positions in space, six months apart. As we know the size of the Earth's orbit, geometry allows calculation of the distance to the star.

▲ **fig A** You can observe the principle behind parallax measurements using your fingers to represent near and distant stars.

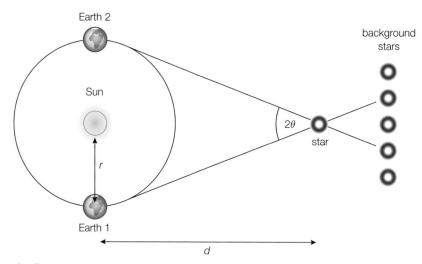

▲ **fig B** Trigonometric parallax measurements.

Using observations of the star to be measured against a background of much more distant stars, we can measure the angle between the star and the Earth in these two different positions in space, six months apart. As we know the size of the Earth's orbit, geometry allows calculation of the distance to the star.

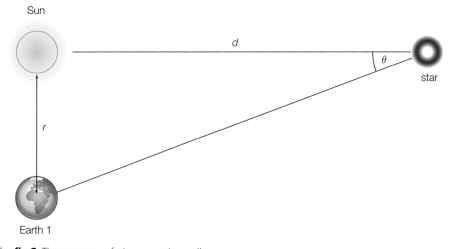

▲ **fig C** The geometry of trigonometric parallax.

By taking the picture of **fig B** and cutting it in half, we get a right-angled triangle formed as shown in **fig C**, with the **parallax angle**, θ.

$$\tan \theta = \frac{r}{d}$$

$$d = \frac{r}{\tan\theta}$$

At small angles, as we are usually using in astronomy, $\tan \theta \approx \theta$. The distance, d, will come out in the same units used to measure r, so we also have:

$$d = \frac{r}{\theta}$$

THE PARSEC

A **parsec (pc)** is a measure of distance. It is an abbreviation of 'parallax second'. It is the distance a star must be from the Sun in order for the angle Earth–star–Sun to be 1 arcsecond.

$$1 \text{ parsec} = \frac{1.5 \times 10^{11}}{\tan (1/3600)}$$

$$= 3.09 \times 10^{16} \text{ m}$$

$$1 \text{ light year} = 3 \times 10^8 \times 365 \times 24 \times 60 \times 60$$

$$= 9.46 \times 10^{15} \text{ m}$$

$$1 \text{ parsec} = 3.27 \text{ light years}$$

The triangle geometry also means that if we have the angle measurement in arcseconds, θ, the distance, d, to the star, measured in parsecs, is easily obtainable from:

$$d = \frac{1}{\theta}$$

WORKED EXAMPLE 1

Astronomers measuring the parallax angle to Alpha Centauri found that the angle measured after a six-month period was 1.52 seconds of arc different from that measured at first. How far is Alpha Centauri from Earth?

$$2\theta = 1.52'' \qquad r = 1.5 \times 10^{11} \text{ m}$$

$$\theta = \frac{1.52}{2} = 0.76''$$

$$= 0.76 \times \frac{1}{3600}$$

$$= 2.11 \times 10^{-4}\,°$$

$$d = \frac{r}{\tan\theta}$$

$$= \frac{1.5 \times 10^{11}}{\tan (2.11 \times 10^{-4})}$$

$$= \frac{1.5 \times 10^{11}}{3.68 \times 10^{-6}}$$

$$d = 4.07 \times 10^{16} \text{ m}$$

Alternatively, straight to parsecs:

$$d = \frac{1}{\theta} = \frac{1}{0.76}$$

$$d = 1.32 \text{ pc}$$

The accuracy of trigonometric parallax depends on the accuracy of the angle measurement. With atmospheric interference for Earth-based telescopes, this was for many years limited to stellar distances of about 100 light years. The European Space Agency's Gaia mission is an orbiting telescope, which has an accuracy in the angle measurement of 24 microarcseconds. This corresponds to allowing accurate distance measurement to stars as far away as 135 000 light years. **Section 11B.4** also highlights how important it is to know stellar distances accurately.

STANDARD CANDLES

▲ **fig D** A light source appears dimmer with increasing distance.

We saw previously that the brightness of a star was linked to its size and its temperature. However, the Stefan–Boltzmann law only considers the power output of the star at its surface. The luminosity of a star is its overall power output over its entire surface. How bright it looks to us, or how much energy we receive from it, depends on how far away the star is (**fig D**). It follows that if we knew the power output of a star, then how bright it appears to us on Earth would give away its distance.

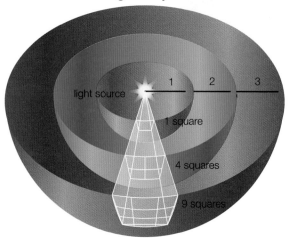

▲ **fig E** The inverse square law, showing energy spread over greater and greater areas with increasing distance from a star.

In **Book 1, Section 3D.4** we saw that measurements of energy intensity are calculated as power per unit area. The general equation we had for intensity was $I = \dfrac{P}{A}$. The inverse square law means that the energy emitted by a star will spread out in all directions over the surface of an increasing sphere (**fig E**). As the surface area of a sphere is $4\pi r^2$, this gives us an equation for the radiant energy intensity at a certain distance, d, from a star:

$$I = \frac{L}{4\pi d^2}$$

where L is the luminosity in watts.

The Sun has a luminosity of $L = 3.8 \times 10^{26}$ W. What is the radiant energy intensity from the Sun at the surface of the Earth?

$$d = 1.5 \times 10^{11}\, \text{m}$$

$$I = \frac{L}{4\pi d^2}$$

$$= \frac{3.8 \times 10^{26}}{4\pi\,(1.5 \times 10^{11})^2}$$

$$I = 1300\ \text{W m}^{-2}$$

Some stars, including some variable stars and supernovae, have properties which mean their luminosity can be determined quite separately from other measurements. These are known as **standard candles**. If we have a figure for the luminosity, and measure the energy intensity (brightness) of the star reaching the Earth, we can then calculate how far away it is by comparing it with a standard candle with the same luminosity.

The luminosity of Betelgeuse is 5.4×10^{30} W. Its radiant energy intensity at the Earth is 1.1×10^{-8} W m^{-2}. How far away is Betelgeuse?

$$I = \frac{L}{4\pi d^2}$$

$$d^2 = \frac{L}{4\pi I}$$

$$d = \sqrt{\frac{L}{4\pi I}}$$

$$= \sqrt{\frac{5.4 \times 10^{30}}{4\pi\,(1.1 \times 10^{-8})}}$$

$$d = 6.2 \times 10^{18}\, \text{m}$$

or: $d = 660\, \text{ly}$

or: $d = 200\, \text{pc}$

VARIABLE STARS

Over a period of years at the beginning of the twentieth century, Henrietta Leavitt, working at the Harvard College Observatory, catalogued many stars in the nearby Magellanic Clouds. She monitored them over time and found that, for some stars, their brightness changed, changing in a repeating cycle. The time period of this oscillation in brightness was constant and, importantly, was in direct proportion to the luminosity of each star (**fig F**).

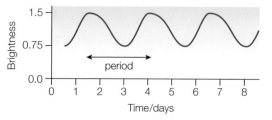

▲ **fig F** The brightness variation typical of RR Lyrae variable stars. These are stars which have left the main sequence, but have not yet exhausted all of their nuclear fuels. They alternately expand and contract causing their light to change repeatedly.

It was possible to calculate the intrinsic luminosity of these stars, as they were close enough to use trigonometric parallax to find their distance. Leavitt had discovered the period–luminosity relation. A longer time period for oscillation meant a brighter star. Astronomers then took this relationship and used it to determine the luminosity of variable stars at much greater distances. From the luminosity, the distance to these stars can be determined using our expression for the radiant energy flux (brightness) observed on Earth. RR Lyrae variable stars allow us to measure distances to about 760 000 parsecs. However, Leavitt particularly studied Cepheid variable stars. There are two types of these, and the more luminous Type I Cepheids give us the greatest distance measurements using the standard candle technique, out to about 40 million parsecs.

TEMPERATURE–LUMINOSITY RELATIONSHIP

One of the simplest methods of determining the luminosity of a star is to look at its spectrum. The peak wavelength gives the temperature from Wien's law, and the width of spectral lines can determine whether or not it is a main sequence star. If it is, and you find its place on the main sequence of the H–R diagram as shown in **fig G**, you can read the luminosity from the y-axis. However, this is one of the least reliable standard candle measurements.

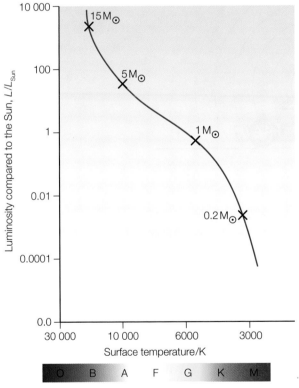

▲ **fig G** The relationship between stellar size and temperature gives luminosity, which in turn can give distance.

CHECKPOINT

SKILLS ADAPTIVE LEARNING

1. Using the data shown in Worked Example 1 on page 189, calculate how far away Alpha Centauri is in astronomical units.

2. The parallax angle to Barnard's Star is 0.545″. How far away is Barnard's Star in
   (a) metres?
   (b) light years?
   (c) parsecs?
   (d) astronomical units?

3. The luminosity of Rigel is 3.9×10^{31} W. At the Earth, Rigel's radiant energy intensity is 5.42×10^{-8} W. How far away is the star?

4. Jupiter's orbit is 5.2 AU from the Sun. If Jupiter is considered as a black body with a temperature of 110 K, and radius 69 900 km, calculate:
   (a) the maximum radiant energy intensity at the surface of the Earth due to Jupiter's electromagnetic emissions (*hint*: you will first need to calculate the luminosity).
   (b) the peak wavelength of Jupiter's energy output.

5. ▶ How does the energy we receive from Jupiter compare with what we get from the Sun?

SUBJECT VOCABULARY

light year the distance that light can travel in one year, in a vacuum, which is about 10^{16} m

astronomical unit (AU) a distance unit, equal to the radius of the Earth's orbit around the Sun: 1 AU = 1.5×10^{11} m

trigonometric parallax a method for measuring the distance to relatively close stars. It works by comparing their measured angle in the sky at six monthly intervals, and using these measurements to calculate the distance using trigonometry

parallax angle the difference in angular observation of a given star, for use in the trigonometric parallax method of measuring the distance to a star

parsec (pc) the distance a star must be from the Sun in order for the parallax angle Earth-star-Sun to be 1 arcsecond: 1 pc = 3.09×10^{16} m

standard candles stars with properties such that their luminosity can be determined from measurements other than brightness

LEARNING OBJECTIVES

■ Define the Doppler effect and explain its application to the light from other stars and galaxies.

■ Use the equation for red shift of light, $z = \dfrac{\Delta\lambda}{\lambda} \approx \dfrac{\Delta f}{f} \approx \dfrac{v}{c}$.

■ Describe Hubble's law and the equation $v = H_0 d$ for objects at cosmological distances.

THE DOPPLER EFFECT

Astronomers first began to look at the spectra of stars in other galaxies during the 1920s. They noticed that the spectra looked very similar to the spectra from stars in our own galaxy but that all the features present were shifted by the same relative amount towards the red end of the spectrum. This phenomenon became known as the **red shift**. This shift is due to the relative motion of other galaxies with respect to ours, in an effect called the **Doppler effect**: an observer receiving waves emitted from a moving body observes that the wavelength of the waves has been altered to a new wavelength.

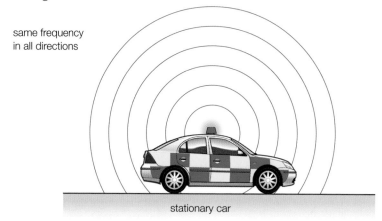

same frequency in all directions

stationary car

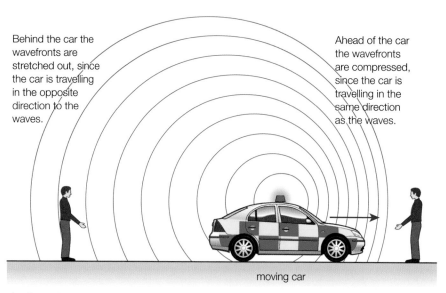

Behind the car the wavefronts are stretched out, since the car is travelling in the opposite direction to the waves.

Ahead of the car the wavefronts are compressed, since the car is travelling in the same direction as the waves.

moving car

▲ **fig A** The Doppler effect causes a change in frequency and wavelength if there is relative motion between the wave source and the observer.

A star or galaxy moving away emits light that appears to have a longer wavelength than expected. We also experience this effect when, for example, we hear a car coming towards us and driving past at a steady rate (**fig A**). As it approaches, the note of its engine rises to a maximum pitch, and then falls as the car travels away. You could imagine the waves getting squashed closer together (shorter wavelength) as the car drives towards you, and then stretched further apart (longer wavelength) as it drives away.

laboratory spectrum

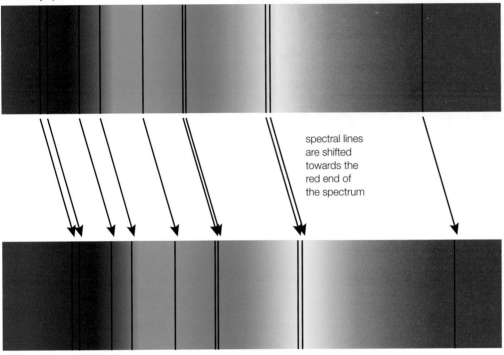

spectral lines are shifted towards the red end of the spectrum

spectrum from a distant galaxy

▲ **fig B** Comparison of light from distant galaxies with light produced in (stationary) Earth-based experiments can allow calculation of the galaxy's speed away from Earth.

LEARNING TIP

Sometimes we see a blue-shift, but this is not common. This shows the star or galaxy is coming towards us.

The amount of red shift a galaxy exhibits, z, allows us to calculate how fast it is moving. This can be done using measurements of either wavelength or frequency changes (**fig B**).

$$z = \frac{\Delta\lambda}{\lambda} \approx \frac{\Delta f}{f} \approx \frac{v}{c}$$

WORKED EXAMPLE 1

In a laboratory sample, the hydrogen alpha spectral absorption line is at a wavelength of 656.285 nanometres. In the spectrum from a nearby star, this line is observed at a wavelength of 656.315 nm. How fast is this star moving and in which direction?

The wavelength from the star is longer than it should be, so the star is moving away.

$\Delta\lambda = 656.315 - 656.285 = 0.030$ nm

$\dfrac{\Delta\lambda}{\lambda} = \dfrac{v}{c}$

$v = c \times \dfrac{\Delta\lambda}{\lambda}$

$\quad = 3 \times 10^8 \times \dfrac{0.030}{656.285}$

$\quad = 13\,700$ m s^{-1}

$v = 13.7$ km s^{-1}

▲ **fig C** This image is a composite taken with Hubble's WFC 3 and ACS on 5 October and 29 November 2011. It shows dwarf galaxy MACS0647-JD, which is one of the most distant objects yet observed. It has a red shift of about $z = 11$.

HUBBLE'S LAW

Astronomers quickly realised that red shift implied that galaxies surrounding us were travelling away from us. In 1929 the American astronomer Edwin Hubble published his finding that the value of a galaxy's red shift is proportional to its distance from us – that is, the further away a galaxy is, the faster it is moving. Hubble's paper had the same effect on the twentieth century view of the Universe as Galileo's work on the solar system had some 300 years earlier. Instead of being static, the Universe was expanding. The philosophical implications continue to intrigue scientists and religious scholars (see **Section 11B.5**).

LEARNING TIP

Local motions deviating from the overall expansion can affect the red shift in a confusing way, which is the main reason that the points in **fig D** show a significant scatter around the best-fit line. Most of the movement causing red shift comes from the expansion of the Universe.

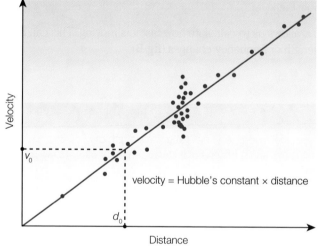

velocity = Hubble's constant × distance

▲ **fig D** As almost all galaxies show red shifts in their spectra, Hubble concluded that all the galaxies must be moving apart from each other and the Universe is expanding.

Considering the speeds of galaxies and their distances, there is a simple relationship between them. This is known as **Hubble's law**:

$$v = H_0 d$$

So, the velocity of a galaxy is directly proportional to the distance to it. We can find the constant of proportionality, the **Hubble constant**, from the gradient of the graph in **fig D**. This has had many values over the years, which demonstrates the immense difficulties involved in accurately determining astronomical distances. Since the launch of the ESA Planck Surveyor, the uncertainty in measurements and so the uncertainty in the Hubble constant has reduced significantly, and data from the Gaia survey will improve it even more. The current value is considered accurate to within

1% and is $H_0 = 70.9 \, \text{km s}^{-1} \, \text{Mpc}^{-1}$, although the most recent data give a slightly lower value (see **fig A** in **Section 11B.5**). With an accurate value for H_0, astronomers can now also use Hubble's law to determine distances to newly observed objects, such as that in **fig C**.

WORKED EXAMPLE 2

A supernova appears in the night sky, and astronomers find that it has a red shift of $z = 0.45$. How far away is the supernova?

$$z = \frac{v}{c}$$
$$v = z \times c$$
$$= 0.45 \times 3 \times 10^8$$
$$= 1.35 \times 10^8 \, \text{m s}^{-1}$$
$$= 1.35 \times 10^5 \, \text{km s}^{-1}$$

According to Hubble's law:

$$v = H_0 d$$
$$d = \frac{v}{H_0}$$
$$= \frac{1.35 \times 10^5}{70.9}$$
$$d = 1900 \, \text{Mpc}$$

Note that the speed was found in km s^{-1}, as we have the Hubble constant in $\text{km s}^{-1} \, \text{Mpc}^{-1}$.

HOW OLD IS THE UNIVERSE?

All distant objects show a red shift and so they are all moving away from us. This implies that the Universe as a whole is expanding. If we imagine time running backwards from the present, then the known Universe would contract back to a point where everything is in the same place. This would be the time of the **Big Bang**, when everything first exploded outwards from that single point. Therefore, if we can find the Hubble constant, it will tell us how quickly the Universe is expanding. From this we can work out when it all started.

For an object to travel a distance d_0 from the beginning of time, at a speed of v_0, the time taken, T_0, can be calculated from the basic equation for speed:

$$\text{speed} = \frac{\text{distance}}{\text{time}}$$
$$v_0 = \frac{d_0}{T_0}$$
$$T_0 = \frac{d_0}{v_0}$$

If we consider the gradient of the Hubble graph, $H_0 = \frac{v_0}{d_0}$.

$$\therefore \quad T_0 = \frac{1}{H_0}$$

Note that in this calculation you should use the same units for the distance and for the length component of the units for **recession velocity**. Usually, though, the Hubble constant is quoted in units of $\text{km s}^{-1} \, \text{Mpc}^{-1}$.

$$H_0 = 70.9 \, \text{km s}^{-1} \, \text{Mpc}^{-1} = 70\,900 \, \text{m s}^{-1} \, \text{Mpc}^{-1}$$
$$1 \, \text{pc} = 3.09 \times 10^{16} \, \text{m}$$
$$1 \, \text{Mpc} = 3.09 \times 10^{22} \, \text{m}$$

$$\therefore \quad H_0 = \frac{70\,900}{3.09 \times 10^{22}}$$
$$= 2.29 \times 10^{-18} \, \text{m s}^{-1} \, \text{m}^{-1}$$
$$H_0 = 2.29 \times 10^{-18} \, \text{s}^{-1}$$
$$T_0 = \frac{1}{H_0}$$
$$\therefore \quad T_0 = \frac{1}{2.29 \times 10^{-18}}$$
$$T_0 = 4.36 \times 10^{17} \, \text{s}$$

This value for T_0 gives the age of the Universe as 13.8 billion years.

CHECKPOINT

SKILLS | CONTINUOUS LEARNING, PERSONAL AND SOCIAL RESPONSIBILITY

1. ▶ Edwin Hubble discovered that the light from other galaxies is Doppler shifted towards the red end of the light spectrum. What does this tell us about the movement of galaxies? Why did this lead Hubble to conclude that the Universe is expanding?

2. (a) Why do spectators at a motor race hear the characteristic 'Neeeoooww' sound as each car passes?

   (b) Why do the drivers hear a sound of constant pitch?

3. The galaxy NGC 7320C has a red shift value of $z = 0.02$.

   (a) At what wavelength would you expect to find the hydrogen alpha line in the spectrum of light from NGC 7320C?

   (b) Use the best modern value for Hubble's constant to calculate the distance to NGC 7320C in megaparsecs.

SKILLS | ANALYSIS

4. ▶ The value for the Hubble constant has varied over the years, from $50 \, \text{km s}^{-1} \, \text{Mpc}^{-1}$ to $100 \, \text{km s}^{-1} \, \text{Mpc}^{-1}$, as different star data were used to draw the graph. Calculate the range of ages of the Universe that this represents.

SUBJECT VOCABULARY

red shift the apparent change in wavelength of a star's spectrum, caused by increasing separation between the star and Earth
Doppler effect the effect that an observer who is receiving waves emitted from a moving body observes that the wavelength of the waves has been altered, to a new wavelength, as a consequence of the relative motion
Hubble's law the recession velocity of a galaxy is directly proportional to the distance to it
Hubble constant the constant of proportionality in Hubble's law
Big Bang Theory the theory that the Universe expanded outwards, from a single point to the currently observed situation
recession velocity the speed at which one object is moving away from another object

LEARNING OBJECTIVES

■ Discuss the controversy over the age and ultimate fate of the Universe associated with the value of the Hubble constant and the possible existence of dark matter.

THE HUBBLE CONSTANT

In **Section 11B.4** we saw that there might be a variation in values of the Hubble constant depending on the observational methods used to find it. This is a very significant and important problem, as it can give a huge variation in possible ages for our Universe. Over the years, the Universe's age based on Hubble's law calculations has given answers ranging from 10 billion to 20 billion years. **Table A** shows a 2014 summary of recent experimental conclusions that indicate a slightly lower value for H_0 than the currently accepted $70.9 \, \text{km s}^{-1}\text{Mpc}^{-1}$. Further observations are needed to confirm these newer results before scientists will agree to change H_0. The result of all of this is that we do not have a fixed answer for the age of our Universe, but this keeps changing as experimental methods improve.

Data set	H_0	Combined results						
	$\text{km s}^{-1}\text{Mpc}^{-1}$	A	B	C	D	E	F	G
WMAP9/2013[a]	70.0 ± 2.2			✓	✓	✓		
WMAP9/2014[b]	69.4 ± 2.2						✓	✓
Planck+WMAPPol/2013[c]	67.3 ± 1.2	✓	✓					
ACT+SPT/2012[d]				✓	✓	✓		
ACT+SPT/2013[d]		✓	✓					
ACT+SPT/2014[d]								✓
BAO/2012[e]				✓	✓	✓		
BAO/2013[e]		✓	✓					
BAO/2014[e]							✓	✓
H_0, Reiss et al. (2011)	73.8 ± 2.4				✓			
H_0, Reiss (2014)	73.0 ± 2.4	✓				✓	✓	✓
Combined H_0 ($\text{km s}^{-1}\text{Mpc}^{-1}$)		67.8	68.3	68.8	69.3	69.2	69.3	69.6
Combined σ ($\text{km s}^{-1}\text{Mpc}^{-1}$)		0.8	0.7	0.8	0.8	0.8	0.8	0.7

table A The values of H_0 derived from different measurement techniques and data sets. From *The 1% concordance Hubble constant*, by Bennett, Larson, Weiland and Hinshaw, Johns Hopkins University, citation: *The Astrophysical Journal*, Volume 794, Number 2 October 2014

FUTURE UNIVERSE?

The Universe is expanding. Will this change in the future? The answer to this question depends critically on the mass of the Universe, and more specifically on the density of matter. Gravity is the force that could slow the expansion down to a stop, and possibly even then start to cause the Universe to contract back inwards (**fig A**). This might eventually end in a **Big Crunch**. The force of gravity between particles decreases with the square of the distance. This means that, if the matter in the Universe is low density, then the gravitational forces between particles, stars and galaxies will be generally weak. The Universe will continue expanding forever. However, if the matter in the Universe has more than a **critical density** then the gravitational forces will succeed in causing the Big Crunch.

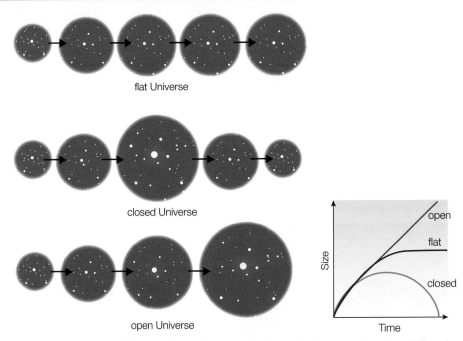

flat Universe

closed Universe

open Universe

▲ **fig A** Our Universe could expand forever, keep slowing down but never reaching a standstill, or slow down to a standstill and start collapsing.

DARK MATTER

To determine the future of the Universe, scientists have been trying to work out its density. However, they have hit upon some big problems. In the 1960s and 1970s, American astronomer Vera Rubin made observations of many galaxies and found a very unexpected result. The stars in the galaxies usually rotate at similar speeds, whatever their distance from the galactic centre. This is completely different to the Solar System, in which the further planets from the Sun orbit much more slowly than those close in.

Galaxies emit an amount of light. We can measure this by measuring their apparent brightness and calculating the inverse square decrease in this which is caused by the distance to the galaxy. From the actual luminosity of a galaxy, we can estimate the mass of all its stars. Galaxies rotate and this means that all the stars they contain must be experiencing a centripetal force towards the centre of the galaxy's rotation. When astronomers measure the rotational speed of the stars in the galaxies, they find that the mass suggested by the luminosity calculations is not nearly enough to create the centripetal force needed to keep the galaxy spinning (**fig C**). In fact, the mass of the stars is generally only about 10% of that needed. This suggests that galaxies must contain a lot of mass that does not emit light. This has been called **dark matter**.

▲ **fig B** Vera Rubin's discovery of unexpected orbital speeds for stars in the Andromeda galaxy highlighted profound problems in the theories about the structure of the Universe.

▲ **fig C** The mass of stars lighting up a galaxy is not nearly enough to provide a gravitational centripetal force strong enough to hold the stars in place against their rotational velocity.

Astronomers have not yet discovered what the dark matter could be. Most stars will have a planetary system, but the mass of all these planets is not nearly enough to hold a galaxy together. Black holes or interstellar gases are good candidates for dark matter, but these (and other similar suggestions) are also not observed in sufficient quantities. We can account for some dark matter with straightforward things such as these, but scientists are convinced that the majority of dark matter is of a different but as yet unknown form.

GRAVITATIONAL LENSING

Einstein's theory of gravity explains that large masses will deform space-time – the fabric of the Universe – so that it accelerates things. This is quite different from Newton's explanation of gravity as a force. The curving of space-time predicts that the direction of travel of light will be affected. Very large masses, such as galaxies or black holes, will cause a bending of light in a similar way as passing through a lens. This effect has been widely observed in photographs of deep space (**fig D**).

▲ **fig D** Gravitational lensing shows the extended arc beneath the Abell 383 galaxy cluster. Such arcs are images of much more distant galaxies where the light has been bent around the Abell 383 galaxy cluster by its enormous mass.

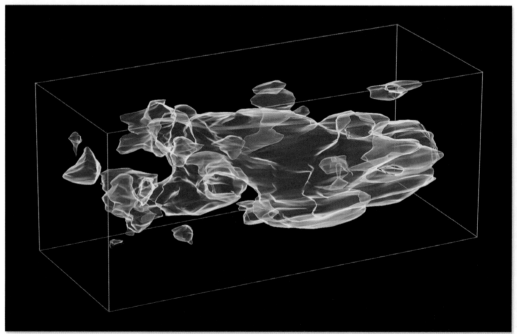

▲ **fig E** Using gravitational lensing, astronomers have verified the existence of dark matter, to the extent that they have made incredibly detailed maps of its locations in space. This picture was generated from images taken by NASA's Hubble Space Telescope.

DARK ENERGY

Until the very end of the twentieth century, it was not known which of the three possible outcomes for the Universe (**fig A**) would happen.

The Hubble Space Telescope was launched in 1990 and its observations of distant supernovae have shown that the expansion of the Universe is not slowing, as we would expect from the gravitational attractions of all the matter in it, but it is actually accelerating. The Universe is getting larger at a faster and faster rate. We do not yet know what is causing this cosmic acceleration, but it is called **dark energy**. All we know about dark energy is how much it is affecting the expansion of the Universe, and so what proportion of the Universe it is. It turns out that dark energy is approximately two-thirds of everything that exists; and we have no idea what it is or how to observe it.

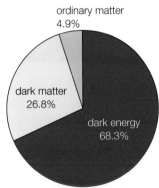

ordinary matter
4.9%

dark matter
26.8%

dark energy
68.3%

▲ **fig F** The ordinary matter we can see and feel makes up only about 5% of the Universe. Dark matter is a little over a quarter of everything, but dark energy is more than two-thirds of the Universe.

CHECKPOINT

SKILLS ▷ REASONING/ARGUMENTATION

1. ▶ Give two reasons why the exact value of the Hubble constant is difficult to determine.

SKILLS ▷ CONTINUOUS LEARNING

2. ▶ What is the importance of the 'critical density' of the Universe?

SKILLS ▷ PERSONAL AND SOCIAL RESPONSIBILITY

3. ▶ Describe two phenomena that have led scientists to believe in the existence of dark matter.

SUBJECT VOCABULARY

critical density the density of matter in the Universe, below which the Universe will keep expanding forever
Big Crunch a possible future for the Universe, in which the Universe contracts inwards to a single point
dark matter a material in the Universe that explains some anomalous behaviours of galaxies; as yet its nature has not been identified
dark energy a source of a force that is causing the expansion of the Universe to accelerate; its nature has not been identified

DARK ENERGY

The European Space Agency (ESA) has a space telescope mission called Euclid, for the study of dark energy, gravity and dark matter. These are high-level physics concepts and are not often explained simply for children. The webpage shown here comes from a website called Our Universe for Kids.

WEBSITE FOR CHILDREN

OUR EXPANDING UNIVERSE

Dark energy and dark matter are not the same. One is matter and the other is energy, but they are both dark, and just like dark matter while it's believed that dark energy is there and does exist, as its believed that there is some evidence of this, the energy cannot be completely proven and confirmed.

Gravity plays an important part in the existence of dark energy and also dark matter, and that is how this phenomenon is measured.

Scientists believe that dark matter makes up about 25% of the universe, with dark energy making up all but the rest at about 70%. That means that only 5% of the visual matter such as the stars, planets and gas clouds makes up the entire universe. That's pretty amazing when you think about it.

So, dark energy is believed to make up around two thirds of basically everything in the universe and we still know almost nothing about it.

Again just like dark matter, we know almost for sure that it's there, and in time we will figure out more about this type of energy, that's for sure. It might turn out to be somewhat different, but we are on the right track.

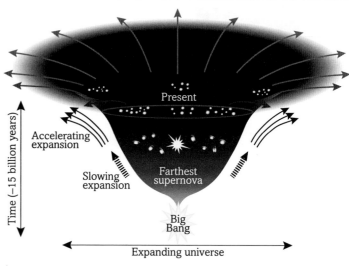

▲ **fig A** Dark matter.

Q&A CORNER

Q: Are dark energy & dark matter the same?

A: No, they are different. One is matter and the other is energy.

Q: How much of the universe is made up of dark energy?

A: $\frac{2}{3}$'s.

Q: What is dark energy?

A: Dark energy is the driving force that expands space.

Q: How much of a percentage of the universe do scientists believe dark energy takes up?

A: 70%

SCIENCE COMMUNICATION

The article is from the website written by a father and his young son.

1 (a) Explain why it is written as a series of short paragraphs.

    (b) Explain why the article is only 213 words long.

    (c) Comment on the size and location of the picture accompanying the article.

2 Describe the level of science presented in this article, and explain how scientific concepts are presented for the intended audience.

INTERPRETATION NOTE

Our Universe for Kids website's stated mission is to: 'Continue to research and create content on different aspects of the universe for the educational and recreational purposes of all kids everywhere that are as interested as we are in the wonders of our amazing universe.'

PHYSICS IN DETAIL

Now we will look at the physics in detail. Some of these questions will link to topics elsewhere in this book, so you may need to combine concepts from different areas of physics to work out the answers.

3 Write a paragraph to explain how 'Gravity plays an important part in ... how [dark matter] is measured'.

4 (a) Explain how astronomers can detect that a galaxy is rotating.

    (b) Explain how unexpectedly fast rotation of galaxies can be explained if there is undetected dark matter mass in the galaxy.

5 Explain the basic difference between dark matter and dark energy.

6 The consortium that is working together to make the ESA Euclid mission happen includes over 120 laboratories from institutions in 15 countries. What would be reasons explaining the need for such a large number of contributors?

PHYSICS TIP

Consider the Doppler effect at points on opposite sides of a rotating galaxy.

ACTIVITY

Write a script for a comparable article for the BBC's international radio station, the BBC World Service. You should include the same main basic information as the kids' website, but should develop the depth of the scientific reporting for an adult public audience, and include some information about the Euclid mission. This task may require some further research.

THINKING BIGGER TIP

Consider that the BBC World Service is broadcast globally and reaches nearly 200 million listeners each week.

Further consider that although only 5% of the population of Earth have English as their first language, your report should be written in English.

[Note: In questions marked with an asterisk (*), marks will be awarded for your ability to structure your answer logically, showing how the points that you make are related or follow on from each other.]

1 What is the peak wavelength of electromagnetic radiation emitted by a star which has a surface temperature of 12 000 K?

A 2.42×10^{-7} m

B 0.242 m

C 34.8 m

D 4.13×10^{6} m [1]

(Total for Question 1 = 1 mark)

2 The intensity of the energy from the Sun arriving at the Earth is 1370 W m^{-2}. The diameter of the Earth's orbit is 3.0×10^{11} m. What is the luminosity of the Sun?

A 4.85×10^{-21} W

B 2.58×10^{15} W

C 3.87×10^{26} W

D 1.55×10^{27} W [1]

(Total for Question 2 = 1 mark)

3 In the laboratory, sodium absorbs light with a wavelength of 5.893×10^{-7} m. The light from a star shows the same sodium absorption line at a wavelength of 5.895×10^{-7} m.

A The star is approaching the Earth at a speed of 0.06 m s^{-1}.

B The star is going away from the Earth at a speed of 0.06 m s^{-1}.

C The star is approaching the Earth at a speed of 100 km s^{-1}.

D The star is going away from the Earth at a speed of 100 km s^{-1}.

[1]

(Total for Question 3 = 1 mark)

4 What do scientists believe will happen at the end of the Universe?

A The Big Bang

B The Big Crunch

C The Big Rip

D Scientists don't have enough evidence to decide. [1]

(Total for Question 4 = 1 mark)

5 The planet Mars has a mean distance from the Sun of 2.3×10^{11} m compared with the Earth's mean distance from the Sun of 1.5×10^{11} m.

(a) Calculate the ratio $\dfrac{\text{Sun's radiation flux at distance of Mars}}{\text{Sun's radiation flux at distance of Earth}}$. [2]

(b) With reference to your answer to (a), comment on the suggestion that Mars could be capable of supporting life. [2]

(Total for Question 5 = 4 marks)

6 Parallax measurements are used to determine the distance to nearby stars, but this method is unsuitable for more distant objects.

Describe how parallax measurements are used to determine the distance to nearby stars and explain how the use of a standard candle enables the distance to more remote objects to be determined. [6]

(Total for Question 6 = 6 marks)

7 The current position of our Sun, S, is shown on the Hertzsprung–Russell (H–R) diagram below.

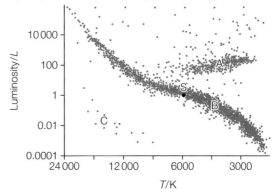

(a) (i) Identify the three main regions of the H–R diagram. [3]

(ii) Add lines to a copy of the diagram to show the evolutionary path of our Sun from the time when it comes to the end of its hydrogen-burning phase. [2]

(b) Most stars are so far way from the Earth that astronomers can only observe them as a point source of radiation.

Explain how astronomers calculate the sizes of these stars, using information from the H–R diagram. [3]

(Total for Question 7 = 8 marks)

8 The graph shows how the velocity varies with distance for several distant galaxies. All the galaxies are moving away from Earth, and there appears to be a linear relationship between the velocity and the distance to the galaxy.

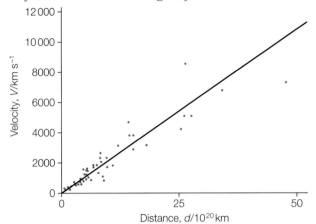

(a) Use the graph to estimate an age for the Universe. [4]

*(b) Describe how astronomers would have determined the velocity of each galaxy. [5]

*(c) Scientists are uncertain about the ultimate fate of the Universe. Explain why. [3]

(Total for Question 8 = 12 marks)

9 A student carries out an experiment on the Stefan–Boltzmann law.
$$L = \sigma T^4 A$$

She uses the filament of a light bulb as a model for a black body radiator.

(a) She obtains the following results.
$$L = 23.5\,\text{W} \pm 2\% \qquad T = 2400\,\text{K} \pm 4\%$$

The student estimates the surface area of the filament A to be $2.0 \times 10^{-5}\,\text{m}^2 \pm 5\%$.

(i) Use her results to calculate an experimental value for the Stefan–Boltzmann constant σ. [1]

(ii) Calculate the percentage uncertainty in the experimental value of σ. [2]

(iii) Calculate the percentage difference between the experimental value of σ and the accepted value, $\sigma = 5.67 \times 10^{-8}\,\text{W m}^{-2}\,\text{K}^{-4}$. [1]

(iv) Use these percentages to comment on the reliability of the experimental value for σ. [1]

(b) The Stefan–Boltzmann law can be written as
$$\ln L = 4\ln T + \ln \sigma A$$

The student obtains a range of values for L and T and plots a graph of $\ln L$ against $\ln T$.

(i) Explain clearly how she could use this graph to obtain a value for σ. [2]

(ii) She realises that she cannot control the temperature of the room.
Explain why this will have almost no effect on the result of the experiment. [1]

(Total for Question 9 = 8 marks)

MATHS SKILLS

For you to be able to develop your skills, knowledge and understanding in Physics, you will also need to develop your mathematical skills in several key areas. This section gives more explanation and examples of some key mathematical concepts you need to understand. Further examples relevant to your International A Level Physics studies are given throughout this book and in Book 1 (IAS).

ARITHMETIC AND NUMERICAL COMPUTATION

USING LOGARITHMS

Many formulae in science and mathematics involve powers. Consider the equation:

$$10^x = 62$$

We know that the value of x lies between 1 and 2, but how can we find a precise answer? The term logarithm means index or power, and logarithms allow us to solve such equations. We can take the 'logarithm base 10' of each side using the **log** button of a calculator.

WORKED EXAMPLE 1

$$10^x = 62$$

$$\log_{10}(10^x) = \log_{10}(62)$$

$$x = 1.792392\ldots$$

We can calculate the logarithm using any number as the base by using the $\log_\square(\square)$ button.

WORKED EXAMPLE 2

$$2^x = 7$$

$$\log_2(2^x) = \log_2(7)$$

$$x = 2.807355\ldots$$

Many equations relating to the natural world involve powers of e. We call these exponentials. The logarithm base e is referred to as the natural logarithm and denoted as **ln**.

There are three laws of logarithms, which are derived from the laws of indices:

$\log a + \log b = \log ab$ (addition rule)

$\log a - \log b = \log a/b$ (subtraction rule)

$\log a^n = n \log a$ (power rule)

WORKED EXAMPLE 3

Using the equation $N = N_0 e^{-\lambda t}$, calculate the decay constant for a substance that has a half-life of 3 years.

$$\frac{N_0}{2} = N_0 e^{-\lambda t_{\frac{1}{2}}}$$

$$\frac{1}{2} = e^{-3\lambda}$$

$$\therefore \quad \ln\left(\frac{1}{2}\right) = -3\lambda$$

$$\lambda = \frac{(\ln 2)}{3} = 0.231\ldots$$

Note that the units for this answer will be years^{-1}.

GRAPHS

USING LOGARITHMIC PLOTS

An earthquake which measures 8.0 on the Richter scale is much more than twice as powerful as an earthquake measuring 4.0 on the Richter scale. This is because the units which are used to measure earthquakes use the concept of logarithm scales in charts and graphs. This helps us to accommodate enormous increases (or decreases) in one variable as another variable changes.

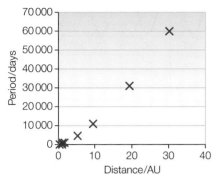 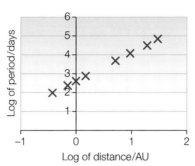

▲ **fig A** These two graphs both show distance from the Sun and period for the eight planets in the Solar System.

If the data being represented follows a power law, using a logarithmic scale will result in a straight-line graph (**fig B**).

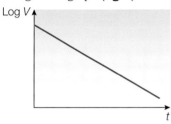

▲ **fig B** We obtain a straight line when we plot log V against time for a discharging capacitor because it is an exponential relationship.

SKETCHING MATHEMATICAL FUNCTIONS

It is helpful to be familiar with standard mathematical functions when modelling physical relationships.

Sine and cosine functions

Remember the following facts about the sine and cosine functions:

- Both functions oscillate between 1 and −1.
- The sine function can be translated 90 degrees to the left to produce the cosine function.
- Each curve repeats itself every 360 degrees (called the period of 360 degrees).
- Each function carries on to positive infinity and minus infinity.

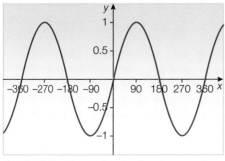 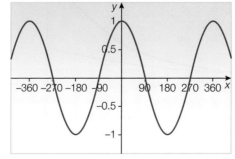

▲ **fig C** The sine and cosine functions: $y = \sin x$ and $y = \cos x$.

The functions $y = \sin^2 x$, $y = \cos^2 x$ are the squares of the sine and cosine functions. Remember that:

• both functions are always positive

• both functions oscillate between 0 and 1.

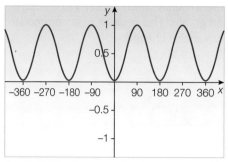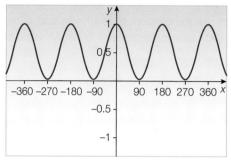

▲ **fig D** The sine squared and cosine squared functions: $y = \sin^2 x$ and $y = \cos^2 x$.

Reciprocal functions

Reciprocal functions such as $y = \dfrac{1}{x}$ and $y = \dfrac{1}{x^2}$ are also useful to recognise. Note that:

• Both functions get closer and closer to zero but never reach zero. The x-axis is referred to as an 'asymptote' – that is, a line that a curve tends towards but never actually reaches.

• The functions are not defined at $x = 0$.

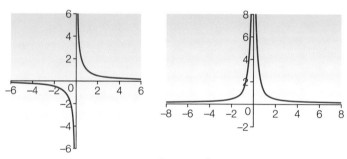

▲ **fig E** Reciprocal functions: $y = \dfrac{1}{x}$ and $y = \dfrac{1}{x^2}$.

Exponential functions

Exponential functions take the form $y = ab^x$. Note that:

• All exponential functions of the form $y = b^x$ pass through the point $(0, 1)$.

• The term exponential is sometimes used to refer to the natural exponential function, which is the special case $y = e^x$.

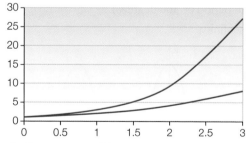

▲ **fig F** Exponential functions: $y = 2^x$ and $y = 3^x$.

THE AREA BETWEEN A CURVE AND THE x-AXIS

The area between a curve and the x-axis was interesting to Sir Isaac Newton and was one of the key concepts that led him to develop the techniques of differentiation and integration. Newton demonstrated that it was possible to find the area between a curve and the axis using an innovative method of summing up strips of area between two distinct limits of the curve.

For many graphs, this area has physical significance:

- The area beneath the curve of a velocity–time graph is equivalent to distance travelled.
- The area under a voltage–charge graph is equivalent to the energy stored for a capacitor.

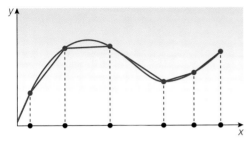

▲ **fig G** The area between the x-axis and the curve can be estimated by summing up the area of strips. Smaller strips give a more accurate answer.

GEOMETRY AND TRIGONOMETRY

USING RADIANS

One theory for how the degree was chosen as a unit of rotation is that Babylonian astronomers used observations of the night sky to estimate an impressively close 360 days as the number of days in a year. A degree was defined as $\frac{1}{360}$ of a full rotation, meaning that a full rotation contains 360 degrees.

When working with trigonometric functions and calculus, it can be much simpler to use radians. A radian is defined as $\frac{1}{2\pi}$ of a full rotation, so a full rotation contains 2π radians. (You can read more about radians in **Section 5B.1**.)

$$2\pi \text{ rad} = 360°$$

$$\pi \text{ rad} = 180°$$

$$1 \text{ rad} \approx 57.296°$$

WORKED EXAMPLE 4

Convert:

(a) 3π radians into degrees

(b) 60° into radians.

(a) 3π radians = 3×180 degrees = 540 degrees

(b) 60 degrees = $60 \left(\frac{\pi}{180}\right)$ radians = $\frac{\pi}{3}$ radians

---

LEARNING TIP

When using a calculator to calculate sin x, cos x and tan x, make sure your calculator is set up to use the angle unit that you need. Most calculators will show 'D' or 'DEG' on the display when in degree mode and 'R' or 'RAD' when in radian mode.

APPLYING YOUR SKILLS

You will often find that you need to use more than one maths technique to answer a question. In this section, we will look at four example questions and consider which maths skills are required and how to apply them.

WORKED EXAMPLE 5

When the switch S is to the left, cell E charges the capacitor C; and when S is moved to the right, C discharges through the ammeter and the resistor R. The discharge current I is given by:

$$I = I_0 \, e^{-t/RC}$$

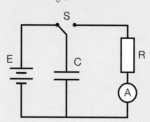

▲ **fig H** Circuit showing charging and discharging capacitor.

(a) Sketch a graph of I against t.

(b) Explain why the area under the graph is equal to the initial charge on the capacitor.

(c) Explain why a graph of $\ln I$ against t will be a straight line.

(d) How would you use a graph of $\ln I$ against t to determine a value for the time constant RC?

(a) You should be able to deduce the shape of the graph from the equation. Note that:

  • There is a finite intercept on the vertical axis.

  • The curve decreases at a decreasing rate.

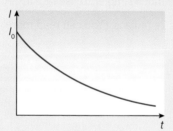

▲ **fig I** Graph showing discharge current against time where $I = I_0 \, e^{-t/RC}$.

(b) From the Formulae sheet, $I = \dfrac{\Delta Q}{\Delta t}$ so $\Delta Q = I \times \Delta t$, which is the area under a graph (**fig I**) of current against time. (To calculate the charge stored you can either perform an integration, which is beyond the scope of the specification, or you can measure the area using the scales on the axes and by counting squares. You will find $Q = EC$.)

(c) Take logarithm base e of the original equation: $I = I_0 \, e^{-t/RC}$

$$\ln I = \ln I_0 - \frac{t}{RC}$$

or:

$$\ln I = -\frac{t}{RC} + \ln I_0$$

This is in the form $y = mx + c$.

So, if we plot $\ln I$ on the y-axis and t on the x-axis, the gradient will be given by $m = -\dfrac{1}{RC}$, and since this is a constant the gradient of the graph will be constant and the line will be straight.

(d) When t is plotted in seconds, the time constant is equal to the gradient of the straight line on the graph of $\ln I$ vs t.

WORKED EXAMPLE 6

An object performs simple harmonic motion. The graph below shows how its velocity varies with time.

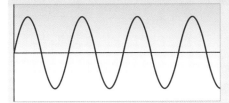

▲ **fig J** Graph showing how velocity varies with time for an object performing simple harmonic motion.

(a) Sketch a graph showing how the acceleration of the object varies with time.

(b) Add to your graph a line showing how the kinetic energy of the body varies with time.

(a) The graph of velocity against time follows the shape of a sine curve.

The graph of acceleration against time follows the shape of a cosine curve.

(b) The graph of kinetic energy against time follows the shape of $y = \sin^2 x$. Note that this does not become negative.

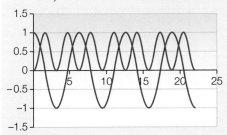

▲ **fig K** Graph showing acceleration and kinetic energy.

WORKED EXAMPLE 7

The graph (**fig L**) shows how the relative intensity of the radiation from a black body varies with temperature. There are three lines on the graph representing the energy from bodies at different temperatures.

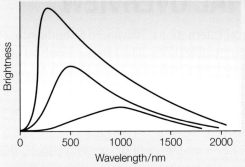

▲ **fig L** Graph showing variation in the relative intensity of the radiation from a black body with wavelength as temperature varies.

(a) Explain which of the lines represents the hottest body.

(b) Explain why the area under the lines increases as shown.

(a) According to Wien's law (from the Formulae sheet: $\lambda_{max}T = 2.898 \times 10^{-3}$ m K), the peak wavelength, λ_{max}, decreases with increasing temperature, so the top line is the hottest body.

(b) The area under the line represents the total energy radiated per unit time or luminosity, which increases with temperature according to the Stefan–Boltzmann law. So, the line showing the highest temperature has the greatest area under it.

WORKED EXAMPLE 8

(a) Explain why the angular speed ω of an orbiting body is given by:

$$\omega = \frac{2\pi}{T}$$

where T is the time taken for one orbit.

(b) Calculate the angular velocity of the Moon around the Earth.

(a) Angular speed $= \dfrac{\text{angular distance}}{\text{time}} = \dfrac{\text{one complete orbit}}{\text{time taken}} = \dfrac{2\pi}{T}$,

where the angular distance for one complete orbit is 2π radians.

(b) For the Moon, $T = 28$ days $= 28 \times 24$ hours $= 28 \times 24 \times 3600$ seconds

So $\omega = \dfrac{2\pi \text{ rad}}{28 \times 24 \times 3600 \text{ s}} = 2.60 \times 10^{-6}$ rad s^{-1}

Since a radian is a ratio of two lengths it has no units and is usually not included in units for angular velocity, so $\omega = 2.60 \times 10^{-6}$ s^{-1}.

PREPARING FOR YOUR EXAMS

IAS AND IAL OVERVIEW

The Pearson Edexcel International Advanced Subsidiary in Physics and the Pearson Edexcel International Advanced Level in Physics are modular qualifications. The International Advanced Subsidiary can be claimed on completion of the International Advanced Subsidiary (IAS) units. The International Advanced Level can be claimed on completion of all the units (IAS and IA2 units).

- International AS students will sit three exam papers. The IAS qualification can be either standalone or contribute 50% of the marks for the International Advanced Level.
- International A Level students will sit six exam papers, the three IAS papers and three IAL papers.

The tables below give details of the exam papers for each qualification.

IAS Papers	Unit 1: Mechanics and Materials	Unit 2: Waves and Electricity*	Unit 3: Practical Skills in Physics I
Topics covered	Topics 1–2	Topics 3–4	Topics 1–4
% of the IAS qualification	40%	40%	20%
Length of exam	1 hour 30 minutes	1 hour 30 minutes	1 hour 20 minutes
Marks available	80 marks	80 marks	50 marks
Question types	short open open-response calculation extended writing	multiple-choice short open open-response calculation extended writing	short open open-response calculation extended writing
Mathematics	For Unit 1 and Unit 2, a minimum of 32 marks will be awarded for mathematics at Level 2 or above. For Unit 3, a minimum of 20 marks will be awarded for mathematics at Level 2 or above.		

* This paper will contain some synoptic questions which require knowledge and understanding from Unit 1.

IAL Papers	Unit 4: Further Mechanics, Fields and Particles**	Unit 5: Thermodynamics, Radiation, Oscillations and Cosmology†	Unit 6: Practical Skills in Physics II
Topics covered	Topics 5–7	Topics 8–11	Topics 5–11
% of the IAL qualification	20%	20%	10%
Length of exam	1 hour 45 minutes	1 hour 45 minutes	1 hours 20 minutes
Marks available	90 marks	90 marks	50 marks
Question types	multiple-choice short open open-response calculation extended writing	multiple-choice short open open-response calculation extended writing	short open open-response calculation extended writing synoptic
Mathematics	For Unit 4 and Unit 5, a minimum of 36 marks will be awarded for mathematics at Level 2 or above. For Unit 6, a minimum of 20 marks will be awarded for mathematics at Level 2 or above.		

** This paper will contain some synoptic questions which require knowledge and understanding from Units 1 and 2.

† This paper will contain some synoptic questions which require knowledge and understanding from Units 1, 2 and 4.

EXAM STRATEGY

ARRIVE EQUIPPED

Make sure you have all of the correct equipment needed for your exam. As a minimum you should take:

- pen (black ink or ball-point pen)
- pencil (HB)
- rule (ideally 30 cm)
- rubber (make sure it's clean and doesn't smudge the pencil marks or rip the paper)
- calculator (scientific).

MAKE SURE YOUR ANSWERS ARE LEGIBLE

Your handwriting does not have to be perfect but the examiner must be able to read it! When you're in a hurry it's easy to write key words that are difficult to understand.

PLAN YOUR TIME

Note how many marks are available on the paper and how many minutes you have to complete it. This will give you an idea of how long to spend on each question. Be sure to leave some time at the end of the exam for checking answers. A rough guide of a minute a mark is a good start, but short answers and multiple choice questions may be quicker. Longer answers might require more time

UNDERSTAND THE QUESTION

Always read the question carefully and spend a few moments working out what you are being asked to do. The command word used will give you an indication of what is required in your answer

Be scientific and accurate, even when writing longer answers. Use the technical terms you've been taught.

Always show your working for any calculations. Marks may be available for individual steps, not just for the final answer. Also, even if you make a calculation error, you may be awarded marks for applying the correct technique.

PLAN YOUR ANSWER

In questions marked with an asterisk (*), marks will be awarded for your ability to structure your answer logically showing how the points that you make are related or follow on from each other where appropriate. Read the question fully and carefully (at least twice!) before beginning your answer.

MAKE THE MOST OF GRAPHS AND DIAGRAMS

Diagrams and sketch graphs can earn marks, often more easily and quickly than written explanations, but they will only earn marks if they are carefully drawn and labelled.

- If you are asked to read a graph, notice the labels and numbers on the x and y axes. Remember that each axis is a number line.
- If asked to draw or sketch a graph, always ensure you use a sensible scale and label both axes with quantities and units.
- If plotting a graph, use a pencil and draw small crosses or dots for the points..
- Diagrams must always be neat, clear and fully labelled.

CHECK YOUR ANSWERS

For open-response and extended writing questions, check the number of marks that are available. If three marks are available, have you made three distinct points?

For calculations, read through each stage of your working. Substituting your final answer into the original question can be a simple way of checking that the final answer is correct. Another simple strategy is to consider whether the answer seems sensible. Be careful to use the correct units.

SAMPLE EXAM ANSWERS

QUESTION TYPE: MULTIPLE CHOICE

Which of the following particles is fundamental?

A electron ☐   B neutron ☐

C pion ☐   D proton ☐   [1]

Question analysis

- Multiple choice questions may require simple recall, as in this case, but sometimes a calculation or some other form of analysis will be required.

- In multiple choice questions you are given the correct answer along with three incorrect answers (called distractors). You need to select the correct answer and put a cross in the box of the letter next to it.

- The three distractors supplied will feature the answers that you are likely to arrive at if you make typical or common errors. For this reason, multiple choice questions aren't as easy as you might at first think. If possible, try to answer the question before you look at any of the answers.

- If you change your mind, put a line through the box with the incorrect answer (⧄) and then mark the box for your new answer with a cross (☒).

- If you have any time left at the end of the paper, multiple choice questions should be put high on your list of priority for checking answers.

Student answer

A electron ☒

This question, in common with all multiple choice questions, only has one correct response. It is often reasonably easy to eliminate some of the responses, but you should only look at the responses when you have thought about your answer first. This question is relatively easy because it tests only one thing – what is meant by the term 'fundamental particle'; however, it requires you to recall the nature of the particles in the responses, so you cannot answer the question without looking at the responses. You can probably recall that protons and neutrons are made up of three quarks so they are not fundamental and can be eliminated. The difficulty arises if you cannot remember that a pion is a meson – a quark and an antiquark.

The student remembers that fundamental means comprised of only one thing, and that the electron is just the electron – with no structure. They can be certain of this and have no need to remember about the pion, since there is only one correct answer.

COMMENTARY

This is a strong answer because:

- The student has not been distracted by response C and has considered all the responses before deciding.

QUESTION TYPE: SHORT OPEN

Show that the kinetic energy of a non-relativistic particle is given by $E_k = \dfrac{p^2}{2m}$ where p is the momentum of the particle and m is its mass. [2]

> This derivation is mentioned in the specification and so is likely to turn up as a question on a paper. Note that for 2 marks you must make two points that are not given in the question.

Question analysis

- Short open questions usually require simple short answers, often one word. Generally, they will require simple recall of the physics you have been taught.

- The command word will tell you what you need to do. When the question is a 'show that' as here, it is important that your answer shows the examiner every step in the derivation (or calculation, if that is what is asked).

- The last step in the derivation is the equation shown in the question (it is also on the Formulae sheet), so you must show two other steps to score the marks.

Student answer

$E_k = \frac{1}{2}mv^2$, where m is mass and v is velocity

Momentum, $p = mv$, so $p^2 = m^2v^2$,

so $\dfrac{p^2}{m} = mv^2$ and $\dfrac{p^2}{2m} = \frac{1}{2}mv^2 = E_k$

> Every step is very clearly shown and in a logical order. Make sure your writing is very clear and easy to read. Take particular care where squares and halves are in the same expression. It does not take much time to write everything down and do it neatly.

COMMENTARY

This is a strong answer because:

- The symbols are explained and there are two steps shown.

- Although the expressions for momentum and for kinetic energy are on the Formulae sheet, it is a good idea to write them down and explain the symbols.

QUESTION TYPE: OPEN RESPONSE

Explain the relationship between path difference and phase difference for two waves.   [4]

Since these terms are not on the Formulae sheet (see Appendix 7 of the specification) there will be one mark each for identifying path difference and phase difference and then two marks for explaining the link.

Question analysis

- The question asks for recall of two terms used in wave analysis, and around a third of the overall marks will be awarded for demonstrating knowledge and understanding.

- There are two marks for stating what the terms mean; there is a mark for identifying the physics relating the terms and then a mark showing how this can be used to answer the question.

- Start by writing down the meaning of the two terms.

Student answer

- Path difference is the difference in the distance travelled by the two waves.
- Phase difference is the difference in the phase of the two waves.
- The number of wavelengths gives the phase difference.

The answer is written as bullet points, which makes it very clear. However, with only three bullet points, it is very unlikely to score all 4 marks. The answer is well structured: it starts with an attempt at explaining the two terms and then tries to explain the connection.

COMMENTARY

This is a weak answer because:

- The path difference is explained in terms of the difference in the distance travelled and gets the first mark.

- The phase difference is not explained using any new words; it simply uses words from the question in a different order, which will not score any marks.

- The connection between the two terms is to do with wavelength, but the answer is too vague about how this connection works. The answer should mention what fraction of a wavelength the path difference leaves.

- A full answer should end with the expression relating the two terms, such as:

$$\text{phase difference} = 2\pi \times \left(\frac{\text{path difference}}{\text{wavelength}}\right)$$

QUESTION TYPE: EXTENDED WRITING

An electrical generator produces electromotive forces from movement. It can consist of a permanent magnet and a coil of wire.

*Use the equation that combines Faraday's law and Lenz's law to explain the factors that affect how a generator produces electromotive forces. [6]

> The question sets out the context for the physics and directs the candidate to writing about magnetic flux and induced e.m.f.s. To explain how a generator works, you need to write out the equation mentioned in the question, and you must refer to it during your answer. Make sure that you get both laws into your answer.

Question analysis

- In questions marked with an asterisk (*), marks will be awarded for your ability to structure your answers logically showing how the points that you make are related or follow on from each other where appropriate.
- Marks will be awarded at different levels, where the highest band (5–6 marks in this case) will only be given if the answer demonstrates comprehensive understanding, uses relevant knowledge of physical concepts and has an explanation that is clear, coherent and logically structured.
- It is vital to plan out your answer before you write it down. There is always space given on an exam paper to do this, so write down the points that you want to make before you answer the question in the space provided. This will help to ensure that your answer is coherent and logical, and that you don't end up contradicting yourself. However, once you have written your answer, go back and draw a line through these notes so that it is clear they do not form part of the answer.

Student answer

A generator uses magnetic flux to produce a voltage. There is a permanent magnet which has its flux linking with the coil of wire. If the coil spins faster, the induced e.m.f. will be larger. This is Faraday's law.

Lenz's law tells us that that any change is opposed, and this is just the conservation of energy in action.

> To score 6 marks for the physics content you must write 6 distinct points. This answer contains five at the most. There is some sort of linkage between the points but it is weak.

COMMENTARY

This is a weak answer because:
- It is important to use A level concepts and terms when answering at this level.
- The question asks for an explanation using the equation but the answer does not do that. Faraday's law should be applied to the changing magnetic flux. There should be a statement about the effect of the number of coils, which should also be explained.
- Whilst the comment about Lenz's law is true, it has no relation to the question.
- There is almost no linkage between the points made.

QUESTION TYPE: CALCULATION

The radiation from a star is observed and measurements are taken.

It is found that the luminosity is 5.20×10^{29} W, and that the peak wavelength in the spectrum of the radiation is 1.71×10^{-7} m.

Calculate the temperature of the star and so calculate a value for the radius of the star. [4]

> This question involves more than one step, so it is important to keep all the information available. Note that the question asks for the radius and not the diameter or area.
>
> Firstly, the question asks for the temperature, and Wien's law will give us this. Then we can use the temperature in the Stefan–Boltzmann law. We can look up the laws on the Formulae sheet (see Appendix 7 of the specification). We can look up the Stefan–Boltzmann constant from the Data sheet (see Appendix 8 of the specification).

Question analysis

- The command word here is 'calculate'. This means that you need to obtain a numerical answer to the question, showing relevant working. If the answer has a unit, this must be included.

- Always try calculation questions. You may get some small part correct that will gain credit.

- The important thing with calculations is to show your working clearly and fully. The correct answer on the line will gain all the available marks; however, an incorrect answer will only lose one mark if your working is shown and is correct. Show the calculation that you are performing at each stage and not just the result. When you have finished, look at your result and see if it is sensible.

- Take an approved calculator into every exam and make sure that you know how to use it!

Student answer

$\lambda_{max} T = 2.898 \times 10^{-3}$ m K, so $T = \dfrac{(2.898 \times 10^{-3} \text{ m K})}{1.71 \times 10^{-7} \text{ m}}$
$= 16\,950$ K $= 1.695 \times 10^{4}$ K

$L = \sigma T^4 A$, so $A = \dfrac{L}{\sigma T^4}$

$= \dfrac{(5.20 \times 10^{29} \text{ W})}{(5.67 \times 10^{-8} \text{ W m}^2 \text{ K}^4 \times (1.695 \times 10^4 \text{ K})^4)} = 1.11 \times 10^{20} \text{ m}^2$

However: $A = 4\pi r^2$ so $r = \left(\dfrac{A}{4\pi}\right)^{1/2} = \left(\dfrac{1.11 \times 10^{20} \text{ m}^2}{4\pi}\right)^{1/2}$
$= 2.97 \times 10^{9}$ m

> Each calculation is laid out clearly on one line and so it is easy to follow where the information goes. Some of the data is to 3 significant figures and some is to 4 significant figures, so the answer should be quoted to 3 significant figures. However, it is sensible to carry 4 significant figures through each part of the calculation, as with the temperature here.

COMMENTARY

This is a strong answer because:

- The use of quantity algebra ensures that the candidate realises that the number at the end of the second line is an area – unit is m^2 – and so cannot be the answer, which is a length, the radius of the star.

- The equation is copied out from the Formulae sheet and then re-arranged. This helps to get the letters in the correct order.

- The numbers are substituted into the re-arranged equation with units.

- The final number is calculated and so is the final unit.

QUESTION TYPE: SYNOPTIC

A nucleus of Radon-226 decays by emitting an alpha particle. When this is observed in a cloud chamber, two tracks are seen travelling in exactly opposite directions.

By considering energy and momentum, describe this decay in as much detail as possible and explain which of the two particles has the most energy. [5]

> The question is open ended in that it asks for as much detail as possible but provides the framework of momentum and energy. The last sentence is in two parts, and it is easy when in an exam to forget about the second part. It is therefore a really good idea to underline key words in the question. Here, the key words might be 'energy and momentum', 'describe' and 'explain'

Question analysis

- A synoptic question is one that addresses ideas from different areas of physics in one context. In your answer, you need to use the different ideas and show how they combine to explain the physics in the context of the question.

- Questions in Paper 3 may draw on any of the topics in this specification.

Student answer

The alpha particle is emitted and the remaining nucleus recoils in the opposite direction because momentum is conserved. No other particle is emitted since there is no momentum perpendicular to the tracks. The energy comes from the mass of the radon nucleus.

The two particles have the same size of momentum and the alpha particle will have more velocity, because it has a smaller mass. As kinetic energy depends on velocity squared, the alpha particle will have more kinetic energy.

> The question has two parts and the answer has two clear sections as well. It is a good idea to lay out your answer to mirror the structure of the question.
>
> The answer uses the terms required as well as other physics terms. It is important that an answer does this and at the appropriate level, so that A level questions get International A Level answers.
>
> The answer would be helped by using some algebraic expressions.

COMMENTARY

This is a fairly weak answer because:

- The answer uses conservation of momentum quite well in the first part, especially since momentum is a vector and travelling in a straight line indicates no 'sideways' momentum and therefore no other particle.

- There is a weak link between the mass and the energy produced in the decay, but it is not developed.

- Conservation of momentum is cited but not really used in the explanation, and the fact that the initial momentum is zero is not used.

- The link between mass and velocity is weak, and as it is the key to the second part this will not score full marks.

COMMAND WORDS

The following table lists the command words used across the IAS/IAL Science qualifications in the external assessments. You should make sure you understand what is required when these words are used in questions in the exam.

COMMAND WORD	THIS TYPE OF QUESTION WILL REQUIRE STUDENTS TO:
ADD/LABEL	Requires adding to or labelling stimulus material given in the question, for example, labelling a diagram or adding units to a table.
ASSESS	Give careful consideration to all the factors or events that apply and identify which are the most important or relevant. Make a judgement on the importance of something, and come to a conclusion where needed.
CALCULATE	Obtain a numerical answer, showing relevant working. If the answer has a unit, this must be included.
COMMENT ON	Requires the synthesis of a number of factors from data/information to form a judgement. More than two factors need to be synthesised.
COMPARE AND CONTRAST	Looking for the similarities **and** differences of two (or more) things. Should not require the drawing of a conclusion. Answer must relate to both (or all) things mentioned in the question. The answer must include at least one similarity and one difference.
COMPLETE/RECORD	Requires the completion of a table/diagram/equation.
CRITICISE	Inspect a set of data, an experimental plan or a scientific statement and consider the elements. Look at the merits and/or faults of the information presented and back judgements made.
DEDUCE	Draw/reach conclusion(s) from the information provided.
DERIVE	Combine two or more equations or principles to develop a new equation.
DESCRIBE	To give an account of something. Statements in the response need to be developed as they are often linked but do not need to include a justification or reason.
DETERMINE	The answer must have an element which is quantitative from the stimulus provided, or must show how the answer can be reached quantitatively.
DEVISE	Plan or invent a procedure from existing principles/ideas.
DISCUSS	Identify the issue/situation/problem/argument that is being assessed within the question. Explore all aspects of an issue/situation/problem. Investigate the issue/situation/problem, etc. by reasoning or argument.

COMMAND WORD	THIS TYPE OF QUESTION WILL REQUIRE STUDENTS TO:
DRAW	Produce a diagram either using a ruler or freehand.
ESTIMATE	Give an approximate value for a physical quantity or measurement or uncertainty.
EVALUATE	Review information then bring it together to form a conclusion, drawing on evidence including strengths, weaknesses, alternative actions, relevant data or information. Come to a supported judgement of a subject's qualities and relation to its context.
EXPLAIN	An explanation requires a justification/exemplification of a point. The answer must contain some element of reasoning/justification; this can include mathematical explanations.
GIVE/STATE/NAME	All of these command words are really synonyms. They generally all require recall of one or more pieces of information.
GIVE A REASON/REASONS	When a statement has been made and the requirement is only to give the reasons why.
IDENTIFY	Usually requires some key information to be selected from a given stimulus/resource.
JUSTIFY	Give evidence to support (either the statement given in the question or an earlier answer).
PLOT	Produce a graph by marking points accurately on a grid from data that is provided and then drawing a line of best fit through these points. A suitable scale and appropriately labelled axes must be included if these are not provided in the question.
PREDICT	Give an expected result or outcome.
SHOW THAT	Prove that a numerical figure is as stated in the question. The answer must be to at least one more significant figure than the numerical figure in the question.
SKETCH	Produce a freehand drawing. For a graph, this would need a line and labelled axes with important features indicated; the axes are not scaled.
STATE WHAT IS MEANT BY	When the meaning of a term is expected but there are different ways of how these can be described.
SUGGEST	Use your knowledge and understanding in an unfamiliar context. May include material or ideas that have not been learnt directly from the specification.
WRITE	When the question asks for an equation.

GLOSSARY

absolute temperature a temperature scale that starts at absolute zero

absolute zero the theoretical temperature at which molecules will no longer be moving, all the kinetic energy has been removed

accrete to grow slowly by attracting and joining with many small pieces of rock and dust, due to the force of gravity

activity the number of radioactive decays in unit time

alpha composed of two protons and two neutrons, the same as a helium nucleus

alpha decay the radioactive process in which a particle combination of two protons and two neutrons is ejected from a nucleus

amplitude the maximum displacement from the equilibrium position

angular displacement the vector measurement of the angle through which something has moved

angular velocity, ω the rate at which the angular displacement changes; unit, radians per second

annihilation the phenomenon in which a particle and its anti-matter equivalent are both destroyed simultaneously in a conversion into energy which is carried away by force carrier particles, such as photons

anti-particle has the same mass but all their other properties are opposite to those of the normal matter particle

astronomical unit (AU) a distance unit, equal to the radius of the Earth's orbit around the Sun: 1 AU = 1.5 × 10^{11} m

atomic number (Z) an alternative name for 'proton number'

background radiation low levels of radiation from environmental sources, always present around us

baryon a particle made of a combination of three quarks

baryon number the quantum number for baryons, whereby each proton or neutron (or other baryon) has a value of $B = 1$

beta an electron emitted at high speed from the nucleus when a neutron decays into a proton

beta-minus decay the radioactive process in which a nuclear neutron changes into a proton, and an electron is ejected from the nucleus

Big Bang Theory the theory that the Universe expanded outwards, from a single point to the currently observed situation

Big Crunch a possible future for the Universe, in which the Universe contracts inwards to a single point

binding energy, ΔE the energy used to hold the nucleus together, converted from the mass deficit, following $E = mc^2$. So it is also the energy needed to break a nucleus apart into its individual nucleons

black body radiator a theoretical object, that completely absorbs all radiation that lands on it

black dwarf the final stage of the life cycle of a small mass star, when nuclear fusion has ceased and it has cooled so that it no longer emits visible light

black hole one of the possible conclusions to the life of a large mass star; a region of space–time in which the gravity is so strong that it prevents anything from escaping, including EM radiation

blue supergiant a very large, very hot star, perhaps 25 000 K

Boyle's law for a constant mass of gas at a constant temperature, the pressure exerted by the gas is inversely proportional to the volume it occupies

bubble chamber a particle detection system in which the particles cause bubbles to be created in a superheated liquid, typically hydrogen

capacitance a measure of the capability of a capacitor; the amount of charge stored per unit voltage across the capacitor, measured in farads, F

capacitor an electrical circuit component that stores charge, and so can be used as an energy store

cathode ray a beam of electrons

Celsius scale a scale of temperature with zero degrees at the freezing point of water, and 100 degrees at the boiling point of water

centripetal acceleration, a the acceleration towards the centre of a circle that corresponds to the changes in direction to maintain an object's motion around that circle

centripetal force the resultant force towards the centre of the circle to maintain an object's circular motion

chain reaction a self-sustaining nuclear reaction in which the products from one individual fission reaction go on to trigger one or more further fissions

Charles's law for a constant mass of gas at a constant pressure, the volume occupied by the gas is proportional to its absolute temperature

control rods materials that can absorb neutrons to stop the triggering of further fission reactions, e.g. boron

critical damping when damping is such that the oscillator returns to its equilibrium position in the quickest possible time, without going past that position

critical density the density of matter in the Universe below which the Universe will keep expanding forever

cyclotron a circular machine that accelerates charged particles, usually following a spiral path

damped oscillations work is done on the damping system and energy is dissipated in the damping system with each oscillation, so the amplitude of oscillations decreases

damping the material, or system, causing energy loss during each damped oscillation

dark energy a source of a force that is causing the expansion of the Universe to accelerate; its nature has not been identified

dark matter a material in the Universe that explains some anomalous behaviours of galaxies; as yet its nature has not been identified

decay ('nuclear' or 'radioactive') a process in which a nucleus' structure is changed, usually accompanied by the emission of a particle

decay constant the probability, per second, that a given nucleus will decay

Doppler effect the effect that occurs when an observer who is receiving waves emitted from a moving body observes that the wavelength of the waves has been altered, to a new wavelength, as a consequence of the relative motion

driving frequency the frequency of an external force applied to a system undergoing forced oscillations

elastic collision a collision in which total kinetic energy is conserved

electric field a region of space that will cause charged particles to experience a force

electric field lines imagined areas where the electric field has an effect

electric field strength the force-to-charge ratio for a charged particle experiencing a force due to an electric field

electric potential a measure of possible field energy in a given location within that field; the energy per unit charge at that location

electromagnetic force one of the four fundamental forces, transmitted by photons, acting between objects with charges

electronvolt the amount of energy an electron gains by passing through a voltage of 1 V

$1 \text{ eV} = 1.6 \times 10^{-19} \text{ J}$
$1 \text{ mega electronvolt} = 1 \text{ MeV} = 1.6 \times 10^{-13} \text{ J}$

equation of state the single equation that defines a gas in terms of its pressure, volume, temperature and quantity:
$pV = NkT$

equipotentials positions within a field with zero potential difference between them

exchange bosons particles that enable the transfer of force. Each of the four fundamental forces has its own exchange boson

exponential curves mathematical functions generated by each value being proportional to the value of one variable as the index of a fixed base: $f(x) = b^x$

Faraday's law induced e.m.f. is proportional to the rate of change of flux linkage

fisson larger nuclei are broken up into smaller nuclides, releasing energy

Fleming's left hand rule a rule for determining the direction of the force generated by the motor effect

flux linkage the amount of magnetic flux interacting with a coil of wire

forced oscillation the oscillation of a system under the influence of an external (usually repeatedly applied) force

free oscillation the oscillation of a system, free from the influence of any forces from outside the system

fuel rod a rod containing the fissionable material, e.g. uranium-235

fundamental particles the most basic particles that are not made from anything smaller. These can be combined to create larger particles

fusion small nuclides combine together to make larger nuclei, releasing energy

gamma high energy, high frequency, electromagnetic radiation, emitted from a nuclear radioactive decay

gravitational field a region of spacetime which is curved. This curvature will cause particles to experience an accelerating force

gravitational potential the amount of work done per unit mass to move an object from an infinite distance to that point in the field

graviton the force carrier particle (or exchange boson) for gravity

gravity the weakest of the four fundamental forces, affecting all objects

hadron a particle which can interact via the strong nuclear force

half-life the time taken for half of the nuclei within a sample to decay. Alternatively, the time taken for the activity of a sample of a radioactive nuclide to reduce to half its initial value

Hertzsprung–Russell diagram a plot of stars, showing luminosity (or absolute magnitude) on the y-axis, and temperature (or spectral class) on the x-axis

Hubble constant the constant of proportionality in Hubble's law

Hubble's law the recession velocity of a galaxy is directly proportional to the distance to it

ideal gas a theoretical gas which does not suffer from the real world difficulties that mean real gases do not perfectly follow the gas laws

impulse force acting for a certain time causing a change in an object's momentum
impulse $= F \times \Delta t$

inelastic collision a collision in which total kinetic energy is not conserved

internal energy the sum of the kinetic and potential energies of all the molecules within a given mass of a substance

isotopes atoms of the same element with different numbers of neutrons in the nuclei

kaon a meson created from any combination of an up or down quark/anti-quark and a strange or anti-strange quark

Kelvin scale an absolute temperature scale with each degree the same size as those on the Celsius scale

kinetic theory the idea that consideration of the microscopic movements of particles will predict the macroscopic behaviour of a substance

Lenz's law the direction of an induced e.m.f. is such as to oppose the change creating it

lepton number the quantum number for leptons, whereby each lepton has a value of $L = 1$

leptons the six fundamental particles which do not interact using the strong nuclear force, only the other three fundamental forces

light year the distance that light can travel in one year, in a vacuum, which is about 10^{16} m

linear accelerator a machine which accelerates charged particles along a straight line

luminosity the rate at which energy of all types is radiated by an object in all directions

magnetic field a region of space that will cause a magnetic pole to feel a force

magnetic field strength an alternative phrase for magnetic flux density

magnetic flux an alternative phrase referring to magnetic field lines

magnetic flux density the ratio of magnetic flux to the area it is passing through

main sequence a diagonal line from top left to bottom right of a Hertzsprung–Russell diagram which marks stars that are in a generally stable phase of their existence

main sequence star a stable star whose core performs hydrogen fusion and produces mostly helium

mass defect, Δm an alternative phrase for 'mass deficit'

mass deficit, Δm the difference between the measured mass of a nucleus and the sum total of the masses of its constituent nucleons

mass number (A) an alternative name for the 'nucleon number'

Maxwell–Boltzmann distribution a mathematical function that describes the distribution of energies amongst particles at a given temperature

meson a particle made of a combination of a quark and an anti–quark

moderator a material used in a nuclear reactor to slow fast-neutrons to thermal speeds

mole the SI unit for amount of substance. One mole contains 6.02×10^{23} molecules of that substance

motor effect a wire carrying a current, held within a magnetic field, will experience a force

natural frequency the frequency of oscillation that a system will take if it undergoes free oscillations

neutron number the total number of neutrons within a given nucleus

neutron star one of the possible conclusions to the life of a large mass star; small and very dense, composed of neutrons

nuclear fission a large nucleus breaks up into two smaller nuclei, with the release of some neutrons and energy

nucleon number the total number of all neutrons and protons in a nucleus

nucleons any of the protons and neutrons comprising a nucleus

orbit the curved path that a planet, satellite or similar object takes around another object in space

oscillate to undertake continuously repeated movements

pair production the phenomenon in which a particle and its anti-matter equivalent are both created simultaneously in a conversion from energy

parallax angle the difference in angular observation of a given star, for use in the trigonometric parallax method of measuring the distance to a star

parsec (pc) the distance a star must be from the Sun in order for the parallax angle Earth–star–Sun to be 1 arcsecond

$1\text{ pc} = 3.09 \times 10^{16}\text{ m}$

period the time taken for one complete oscillation

photoelectric effect when electrons are released from a metal surface because the metal is hit by electromagnetic radiation

photons the quantum of electromagnetic radiation, and the force carrier for the electromagnetic force

pion a meson created from any combination of up and down quark/anti-quark pairings

planetary nebula the remains of an explosion at the end of the life cycle of a low-mass star; material which may eventually join together into new planets

plum pudding model a pre-1911 model of the atom, in which the main body of the atom is made of a positively charged material (the pudding 'dough') with electrons (the 'plums') randomly scattered through it

poles the magnetic equivalent of a charge on a particle: north pole or south pole

potential difference the change in potential between two locations in a given field

pressure law for a constant mass of gas at a constant volume, the pressure exerted by the gas is proportional to its absolute temperature

primary coil the first coil in a transformer, through which the supply current passes

proton number the number of protons in the nucleus of an atom

protostar the coalescence of dust and gas under the force of gravity, prior to the start of nuclear fusion in its core, which will go on to become a star

quarks the six fundamental particles that interact with each other using the strong nuclear force (as well as all other forces)

radian a unit of angle measurement, one radian is equivalent to 57.3 degrees

recession velocity the speed at which one object is moving away from another object

red giant a large star which is cooler than our Sun, e.g. 3000 K

red shift the apparent change in wavelength of a star's spectrum, caused by increasing separation between the star and Earth

resonance very large amplitude oscillations that occur when a driving frequency matches the natural frequency of a system

root-mean-square speed the square root of the arithmetic mean value of the squares of the speeds of particles in an ideal gas

secondary coil the second coil in a transformer, through which the output current passes

simple harmonic motion (SHM) the oscillation of a system in which a force is continually trying to return the object to its centre position and this force is proportional to the displacement from that centre position

specific heat capacity the energy required to raise the temperature of one kilogram of a substance by one degree kelvin

specific latent heat the energy required to change the state of one kilogram of a substance at a constant temperature

standard candles stars with properties such that their luminosity can be determined from measurements of brightness

Standard Model the name given to the theory of all the fundamental particles and how they interact. This is the theory that currently has the strongest experimental evidence

Stefan–Boltzmann law the power output from a black body is proportional to its surface area and the fourth power of its temperature in kelvin, $L = \sigma A T^4$

strangeness the quantum number for strange quarks, whereby each one has a value of $S = -1$

strong nuclear force the extremely short-range force between hadrons (such as protons and neutrons)

supernova the explosion of a large mass star at the end of its lifetime, when it becomes extremely unstable

synchrotron a machine that accelerates charged particles around a fixed circular path

tesla (T) the unit for magnetic flux density, or magnetic field strength

thermal neutron a relatively slow-moving neutron

thermionic emission the release of electrons from a metal surface caused by heating of the metal

time constant (for a capacitor-resistor circuit) the product of the capacitance and the resistance, giving a measure of the rate for charging and discharging the capacitor. This is the time taken for the current, voltage or charge of a discharging capacitor to fall to 37% of its initial value. Symbol: tau, τ (sometimes T is used instead)

trigonometric parallax a method for measuring the distance to relatively close stars. It works by comparing their measured angle in the sky at six monthly intervals, and using these measurements to calculate the distance using trigonometry

uniform acceleration acceleration that always has the same value; constant acceleration

weak nuclear force one of the four fundamental forces, transmitted by W or Z bosons, acting at extremely short ranges (10^{-18} m); it can affect all matter particles

weber, Wb the unit of measurement of magnetic flux, Φ, (and magnetic flux linkage, $N\Phi$)

weber-turns the unit for magnetic flux linkage

white dwarf a small, hot star, perhaps 10 000 K

Wien's law the relationship between the peak output wavelength and temperature for a black body radiator is given by the equation: $\lambda_{max}T = 2.898 \times 10^{-3}$ m K

INDEX

Acknowledgements
The author and publisher would like to thank the following individuals and
organisations for permission to reproduce photographs:

(Key: b-bottom; c-centre; l-left; r-right; t-top)

Image Credit(s):
2 Getty Images: Simon Bruty/Stockbyte/Getty Images **4 Shutterstock**: (t) Evgeny
Murtola/Shutterstock; (b) Valdis Torms/Shutterstock **5 Pearson Education Ltd**
8 Alamy Stock Photo: Stocktrek Images, Inc./Alamy Stock Photo **14 Getty
Images**: Emmanuel Reze/EyeEm/Getty Images **16 Shutterstock**: Zhukovvvlad/
Shutterstock **17 Shutterstock:** Gwycech/Shutterstock **19 Shutterstock:** Nicholas
Piccillo **20 Alamy Stock Photo**: Pete Vazquez/RGB Ventures/SuperStock /Alamy
Stock Photo **25 Shutterstock:** Uellue/Shutterstock **26 Alamy Stock Photo**:
Luscious Frames/Alamy Stock Photo **28 Shutterstock:** Tyler Oslon/Shutterstock
30 Science Photo Library: (l) Andrew Lamber Photography/Science Photo Library;
(r) Trevor Clifford Photography/Science Photo Library **31 Caltech Archives** Caltech
Archives **36 Science Photo Library**: Trevor Clifford Photography/Science Photo
Library **38 Shutterstock:** Pan Demin/Shutterstock **42 Getty Images**: Greenlin/
Moment/Getty Images **44 Science Photo Library**: Trevor Clifford Photography/
Science Photo Library **45 Alamy Stock Photo**: Sciencephotos/Alamy Stock
Photo **47 SuperStock:** Transtock/SuperStock **52 AIP Publishing:** Reproduced
from 'Paper-based ultracapacitors with carbon nanotubes-graphene composites',
Journal of Applied Physics, Vol. 115, Issue 16 (Li, J., Cheng, X., Sun, J., Brand,
C., Shashurin, A., Reeves, M. and Keidar, M. 2014), with the permission of AIP
Publishing **56 Shutterstock:** Chaiviewfinder/Shutterstock **58 Alamy Stock Photo**:
Phil Degginger/Alamy Stock Photo **60 Science Photo Library**: Trevor Clifford
Photography/Science Photo Library **61 Science Photo Library**: (l) Trevor Clifford
Photography/Science Photo Library; (r)**University of Texas**: University of Texas/
Mr. Eric E. Zumalt (b) **Zettl Research group** Noah Boadzin/ Zettl Research
group **62 Science Photo Library**: Trevor Clifford Photography/Science Photo
Library **63 Alamy Stock Photo:** Dorling Kindersley ltd/Alamy Stock Photo **65
Science Photo Library**: Trevor Clifford Photography/Science Photo Library **66
Science Photo Library**: Trevor Clifford Photography/Science Photo Library **68
Shutterstock**: Angyalosi Beata/Shutterstock **72 Shutterstock:** Sergey Nivens/
Shutterstock **78 Science Photo Library**: DR DAVID WEXLER, COLOURED BY DR
JEREMY BURGESS/Science Photo Library **84 Shutterstock**: Agnieszka Skalska/
Shutterstock **91 Science Photo Library**: (t) Cern/Science Photo Library; **Getty
Images**: (b) Science & Society Picture Library/Getty Images **92 Science Photo
Library**: (br) Carl Anderson/Science Photo Library **93 Alamy Stock Photo**: Alfred
Abad/Age Fotostock/Alamy Stock Photo **94 Science Photo Library**: (b) Cern/
Science Photo Library **Alamy Stock Photo**: (t) Simon Hadley/Alamy Stock Photo
100 Alamy Stock Photo: Mehau Kulyk/Science Photo Library/Alamy Stock
Photo **102 Alamy Stock Photo**: Granger Historical Picture Archive/Alamy Stock
Photo **103 Lawrence Berkeley Nat'l Lab** Photo courtesy of Berkeley Lab **105
Shutterstock:** Dominionart/Shutterstock **116 Getty Images**: Stephen Frink/Image
Source/Getty Images **118 Alamy Stock Photo**: Chris Hellier/Alamy Stock Photo
121 Science Photo Library: Trevor Clifford Photography/Science Photo Library
125 Alamy Stock Photo: Sciencephotos/Alamy Stock Photo **126 DK Images**:
Andy Crawford/DK Images **128 Ballooning Bradleys** Courtesy of Ballooning
Bradleys **131 Shutterstock:** J.Helgason/Shutterstock **132 Getty Images**: Charissa
Van Straalen/EyeEm/Getty Images **139 Science Photo Library**: Trevor Clifford
Photography/Science Photo Library **145 Alamy Stock Photo:** A. T. Willett/
Alamy Stock Photo **146 Science Photo Library:** Simon Fraser/Medical Physics,
RVI, Newcastle/Science Photo Library **150 Getty Images:** Johannes Hulsch/
EyeEM Premium/Getty Images **158 Shutterstock:** (t) Monkey Business Images/
Shutterstock **123RF:** (b) entropic/123RF **162 Science Photo Library:** (t) Trevor
Clifford Photography/Science Photo Library; **Getty Images:** (br) Steve Bronstein/
The Image Bank/Getty Images; **Alamy Stock Photo:** (bl) DaniFoto/Alamy Stock
Photo **163 Alamy Stock Photo:** Xinhua/Alamy Stock Photo **164 Shutterstock:**
Alvinku/Shutterstock **168 Getty Images:** AlenaPaulus/E+/Getty Images **170
Shutterstock:** (l) Lukasz Pawel Szczepanski/Shutterstock **178 Getty Images:** 2010
Luis Argerich/Moment/Getty Images **180 Shutterstock:** (l) Srifoto/Shutterstock;
NASA: (r) E. Olszewski/NASA **186 Alamy Stock Photo:** NG Images/Alamy Stock
Photo **188 Science Photo Library:** Trevor Clifford Photography/Science Photo
Library **190 Shutterstock:** Ekaterina Garyuk/Shutterstock **194 NASA:** NASA, ESA,
and M. Postman and D. Coe (STScI) and CLASH Team **197 Alamy Stock Photo:**
(r) Walter Oleksy/Alamy Stock Photo; **Shutterstock:** (l) Albert Barr/Shutterstock
198 NASA: (t) NASA Images / ESA, J. Richard (Center for Astronomical Research
/ Observatory of Lyon, France), and J.-P. Kneib (Astrophysical Laboratory of
Marseille, France) **Science Photo Library:** (b) R. Massey, Caltech/ NASA / ESA /
STSCI / Science Photo Library **200 NASA:** Ann Feild/NASA